Lecture Notes in Mathematics 1601

Editors:
A. Dold, Heidelberg
F. Takens, Groningen

T0222676

Alexander Pott

Finite Geometry
and Character Theory

 Springer

Author

Alexander Pott
Institut für Mathematik der Universität Augsburg
Universitätsstraße 14
D-86135 Augsburg, Germany

Mathematics Subject Classification (1991): 05B05, 05B10, 05B20, 05E20, 11R04, 11T24, 20B25, 51E15, 51E30, 94A55, 94B15

ISBN 3-540-59065-X Springer-Verlag Berlin Heidelberg New York

© Springer-Verlag Berlin Heidelberg 1995
Printed in Germany

Typesetting: Camera-ready TeX output by the author
SPIN: 10130289   46/3142-543210 - Printed on acid-free paper

Preface

It is the aim of this monograph to demonstrate the applications of algebraic number theory and character theory in order to solve problems in finite geometry, in particular problems about difference sets and their corresponding codes and sequences. The idea is the following: Assume that D is an element in the integral group ring $\mathbf{Z}[G]$ which satisfies a certain "basic" equation like $DD^{(-1)} = M$. (We will always assume that G is a finite group.) If G is abelian, we can apply characters χ (or 1-dimensional representations) to compute $\chi(M)$. Using results from algebraic number theory, it is then (sometimes) possible to get informations about $\chi(D)$. This might be helpful to <u>construct</u> D or to prove that no such D can exist which satisfies the basic equation (<u>non-existence</u>). These are the two fundamental problems concerning difference sets, and they will be investigated in this monograph. The classical paper where this approach has been first used is Turyn [167].

In the first chapter we will present the algebraic approach to attack these two problems. In the second chapter we will describe several constructions of difference sets and state some non-existence results. We do not present all known results in complete generality. But the proofs of many theorems about difference sets use ideas which are very closely related to the techniques developed in this monograph, in particular those presented in the first two chapters. I hope that these chapters are particularly useful for those readers who want to become acquainted with "difference set problems".

In most cases our groups are abelian. Of course, one can also ask for solutions of the "basic" equation in $\mathbf{Z}[G]$ if G is non-abelian. Sometimes our techniques are also applicable if G has a "large" homomorphic image which is abelian. But even if this is not the case one can try to use the results from algebraic number theory which are so useful in the abelian case: Instead of 1-dimensional representations one has to consider higher-dimensional representations of the group G. The entries in the matrices of these representations are algebraic integers, which satisfy certain equations (similar to the abelian case), although the equations might be slightly more involved. The reader is referred to the papers by Liebler and Smith [121] and Davis and Smith [56], where this approach has been used quite successfully. In these papers, the same results from algebraic number theory are used which have been already used in the abelian case: Although I restrict myself basically to abelian groups, this book will be also useful for those who are more interested in the non-abelian case.

This monograph is not a book about difference sets in the usual sense (like the books by Baumert [26] and Lander [116]). We describe the known series of difference sets, but this part of the monograph (Section 2.1) has the character of a brief survey. A nice survey about difference sets has been written by Jungnickel [100]. After this survey has been published, several new and exciting results about difference sets (in particular about Hadamard or Menon difference sets and about McFarland difference sets) have been proved (which can be found, without proofs, in this monograph). The only difference sets which we study in more detail are the Singer difference sets and their generalizations in Chapter 3.

In the Chapters 4 and 5 we study several interesting classes of so called relative difference sets. The semiregular case in Chapter 4 is somehow the "relative" analogue of the Hadamard difference sets. In a special case, these semiregular difference sets describe projective planes with a quasiregular collineation group. This is one of three cases which we investigate in Chapter 5. The main problem there is to find numerical restrictions on the possible orders of projective planes admitting quasiregular collineation groups. In one of these cases we can show that n has to be a prime power, or n is a square.

Difference sets and their generalizations give rise to sequences with interesting correlation properties. We investigate these sequences in Chapter 6 (Barker sequences, (almost) perfect sequences). Moreover, we determine the dimensions of certain codes related to difference sets. But we investigate sequences related to difference sets not only in this last chapter. For instance, the Gordon-Mills-Welch difference sets and the corresponding sequences are studied in Chapter 3. Therefore, this monograph can be also useful for mathematicians and engineers who are interested in sequences with "nice" autocorrelation properties: The autocorrelation properties can be translated into group ring equations, and we can use all the techniques developed in this book to find solutions of these equations, or we can try to prove that no solutions can exist. Moreover, we can determine the linear complexities of the sequences.

This monograph is not intended to be a text book: There are too many theorems where we do not give proofs but just refer to the original papers. But this monograph should be a good preparation to read more advanced and technical papers about difference sets. I am sure that the book can be very useful in any course on algebraic design theory (or even algebraic coding theory).

I have posed several questions throughout the text hoping that these will stimulate future research.

Finally, I would like to thank all my colleagues with whom I have worked in the past years and who have supported my research: They all have influenced the presentation in this monograph (without knowing it). In particular, I would like to thank my (former) supervisor Dieter Jungnickel. For the careful reading of (parts of) the final version of the manuscript I thank Bernhard Schmidt and Dieter Jungnickel.

I would also like to thank the following institutions: First of all, I have done most of the research which is contained in this monograph at the *Justus-Liebig-Universität Gießen*. Most of the writing of this text has been done at the *Universität Augsburg*, and the final work has been done at the *Gerhard-Mercator-Universität (Gesamthochschule) Duisburg*.

Alexander Pott

Contents

Chapter 1

Preliminaries: Incidence structures with Singer groups

1.1 Basic definitions

An **incidence structure** or a **design** $\mathcal{D}$ is a triple $(\mathcal{V}, \mathcal{B}, I)$ consisting of two disjoint sets $\mathcal{V}$ and $\mathcal{B}$ and a subset $I \subseteq \mathcal{V} \times \mathcal{B}$. Elements in $\mathcal{V}$ are called **points**, those in $\mathcal{B}$ **blocks**. We will always denote points by small and blocks by capital letters. We say that a point p **lies** on a block B, or that a point p is **incident** with B if $(p, B) \in I$. In this monograph, we consider only **finite** incidence structures, i.e. incidence structures where both $\mathcal{V}$ and $\mathcal{B}$ are finite sets. Such a finite incidence structure can be described by an **incidence matrix** $\mathbf{A}$: We label the rows of $\mathbf{A}$ with the points and the columns with the blocks. Then we define the (p, B)-entry of $\mathbf{A}$ by

$$a_{p,B} = \left\{ \begin{array}{ll} 1 & \text{if } (p, B) \in I \\ 0 & \text{otherwise.} \end{array} \right.$$

Since $\mathbf{A}$ is a matrix with entries 0 and 1, we can define $\mathbf{A}$ over any field or any ring which contains an identity element. For the time being, let us think of $\mathbf{A}$ as a matrix over the integers. Then the (p, q)-entry in

$$\mathbf{M} := \mathbf{A}\mathbf{A}^{\mathbf{t}} \tag{1.1}$$

is the number of blocks through p and q (here $\mathbf{A}^t$ is the transpose of $\mathbf{A}$).

Example 1.1.1 *The matrix*

$$\begin{pmatrix} 1 & 1 & 0 & 1 & 0 & 0 & 0 \\ 0 & 1 & 1 & 0 & 1 & 0 & 0 \\ 0 & 0 & 1 & 1 & 0 & 1 & 0 \\ 0 & 0 & 0 & 1 & 1 & 0 & 1 \\ 1 & 0 & 0 & 0 & 1 & 1 & 0 \\ 0 & 1 & 0 & 0 & 0 & 1 & 1 \\ 1 & 0 & 1 & 0 & 0 & 0 & 1 \end{pmatrix}$$

is the incidence matrix of a so called **projective plane of order 2**. *Not surprisingly, this matrix is called* **circulant**, *a term which we will soon make more precise. It is the theme of this monograph to study classes of designs which admit circulant or, more generally, group invariant incidence matrices.*

The definition of a design is rather general, perhaps too general to expect interesting results. Therefore, in most cases we impose some additional structure on the design: Here we will restrict the number of blocks through two distinct points, or, in terms of the incidence matrices, we assume that the matrix **M** in (1.1) is a "nice" matrix. In the example above, we get

$$\mathbf{M} = \begin{pmatrix} 3 & 1 & 1 & 1 & 1 & 1 & 1 \\ 1 & 3 & 1 & 1 & 1 & 1 & 1 \\ 1 & 1 & 3 & 1 & 1 & 1 & 1 \\ 1 & 1 & 1 & 3 & 1 & 1 & 1 \\ 1 & 1 & 1 & 1 & 3 & 1 & 1 \\ 1 & 1 & 1 & 1 & 1 & 3 & 1 \\ 1 & 1 & 1 & 1 & 1 & 1 & 3 \end{pmatrix} = 2\mathbf{I}_7 + \mathbf{J}_7,$$

where $\mathbf{I}_v$ and $\mathbf{J}_v$ denote the identity matrix and the "all-one-matrix" of size v. Thus the projective plane of order 2 is an incidence structure where any two distinct points are joined by exactly one line, and where the line size is three.

Example 1.1.2 Another example of a design where **M** is an "interesting" matrix is the following: We take

$$\mathbf{A} = \begin{pmatrix} 0 & 1 & 1 & 0 & 1 & 0 & 0 & 0 \\ 0 & 0 & 1 & 1 & 0 & 1 & 0 & 0 \\ 0 & 0 & 0 & 1 & 1 & 0 & 1 & 0 \\ 0 & 0 & 0 & 0 & 1 & 1 & 0 & 1 \\ 1 & 0 & 0 & 0 & 0 & 1 & 1 & 0 \\ 0 & 1 & 0 & 0 & 0 & 0 & 1 & 1 \\ 1 & 0 & 1 & 0 & 0 & 0 & 0 & 1 \\ 1 & 1 & 0 & 1 & 0 & 0 & 0 & 0 \end{pmatrix}.$$

Then

$$
\mathbf{A}\mathbf{A}^t = \begin{pmatrix}
3 & 1 & 1 & 1 & 0 & 1 & 1 & 1 \\
1 & 3 & 1 & 1 & 1 & 0 & 1 & 1 \\
1 & 1 & 3 & 1 & 1 & 1 & 0 & 1 \\
1 & 1 & 1 & 3 & 1 & 1 & 1 & 0 \\
0 & 1 & 1 & 1 & 3 & 1 & 1 & 1 \\
1 & 0 & 1 & 1 & 1 & 3 & 1 & 1 \\
1 & 1 & 0 & 1 & 1 & 1 & 3 & 1 \\
1 & 1 & 1 & 0 & 1 & 1 & 1 & 3
\end{pmatrix} = 3\mathbf{I}_8 + \mathbf{J}_8 - \mathbf{J}_2 \otimes \mathbf{I}_4 .
$$

In this example, "$\otimes$" denotes the usual Kronecker product of two matrices: If $\mathbf{A} = (a_{i,j})$ and $\mathbf{B} = (b_{k,l})$ are two matrices with $1 \leq i \leq n$, $1 \leq j \leq m$, $1 \leq k \leq n'$ and $1 \leq l \leq m'$, then $\mathbf{A} \otimes \mathbf{B}$ is an $nn' \times mm'$-matrix

$$
\mathbf{A} \otimes \mathbf{B} := \begin{pmatrix}
a_{1,1}\mathbf{B} & \cdots & a_{1,m}\mathbf{B} \\
\vdots & & \vdots \\
a_{n,1}\mathbf{B} & \cdots & a_{n,m}\mathbf{B}
\end{pmatrix} .
$$

The incidence structure described by the incidence matrix $\mathbf{A}$ in Example 1.1.2 is called a (group) divisible design: The points can be partitioned into 4 "groups" or point classes of size 2 (corresponding to lines i and $i + 4$, $0 \leq i \leq 3$, in $\mathbf{A}$) such that two distinct points in distinct "groups" are joined by exactly one line, whereas points in the same group are not joined by a line. The line size is three, and there are precisely three lines through each point. In general, a **divisible design** with parameters $(m, n, k, \lambda_1, \lambda_2)$ is an incidence structure $\mathcal{D}$ with mn points for which a suitable incidence matrix satisfies the equation

$$
\mathbf{A}\mathbf{A}^t = (r - \lambda_1)\mathbf{I}_{mn} + (\lambda_1 - \lambda_2)\mathbf{J}_n \otimes \mathbf{I}_m + \lambda_2\mathbf{J}_{mn}, \tag{1.2}
$$

provided that each block has exactly k points (in matrix terms $\mathbf{J}_{mn}\mathbf{A} = k\mathbf{J}_{mn}$). In other words, we can partition the points into m point classes or "groups" of size n with the following properties: Two distinct points are joined by exactly λ_1 blocks, if they are in the same point class, otherwise they are joined by λ_2 blocks. The block size is k and there are precisely r blocks through each point. The parameter r can be determined from the following equation which can be proved by an easy counting argument:

$$
r(k - 1) = \lambda_1(n - 1) + \lambda_2(mn - n). \tag{1.3}
$$

Similarly, the total number b of blocks is determined by

$$
vr = bk. \tag{1.4}
$$

The design in Example 1.1.2 is a divisible design with parameters $(4, 2, 3, 0, 1)$. This design has another interesting property, namely, the number v of points equals the number b of blocks, equivalently, the number of blocks through a point equals the block size. Designs with this property are called **square**. Then

(1.3) shows that the five parameters m, n, k, λ_1 and λ_2 are not independent, therefore in the square case we consider actually a 4-parameter family and not a 5-parameter family. The term **symmetric** is reserved for designs whose dual is a design "of the same type", in particular, a design with the same parameters: The **dual** is the incidence structure obtained by interchanging the roles of points and blocks. In matrix language, the incidence matrix of the dual design is the transpose of the incidence matrix of the original design. This is the reason why we denote the dual of $\mathcal{D}$ by $\mathcal{D}^t$. Examples of square designs are known which are not symmetric:

Example 1.1.3 (Connor [47]) The matrix

$$
\mathbf{A} = \begin{pmatrix}
1 & 1 & 1 & 1 & 0 & 0 & 0 & 0 \\
0 & 0 & 0 & 0 & 1 & 1 & 1 & 1 \\
1 & 0 & 0 & 1 & 1 & 0 & 0 & 1 \\
0 & 1 & 1 & 0 & 0 & 1 & 1 & 0 \\
0 & 1 & 0 & 1 & 0 & 1 & 0 & 1 \\
1 & 0 & 1 & 0 & 1 & 0 & 1 & 0 \\
0 & 0 & 1 & 1 & 0 & 0 & 1 & 1 \\
1 & 1 & 0 & 0 & 1 & 1 & 0 & 0
\end{pmatrix}
$$

is the incidence matrix of a $(4,4,4,0,2)$-divisible design $\mathcal{D}$. The dual is not a divisible design, since the block intersection numbers in $\mathcal{D}$ are 1, 2 and 3. Note that the block intersection numbers in $\mathcal{D}$ become the number of blocks through distinct points in $\mathcal{D}^t$.

We can determine the eigenvalues of the matrix on the right hand side of (1.2):

Lemma 1.1.4 *The matrix*

$$(r - \lambda_1)\mathbf{I}_{mn} + (\lambda_1 - \lambda_2)\mathbf{J}_n \otimes \mathbf{I}_m + \lambda_2\mathbf{J}_{mn} \tag{1.5}$$

has eigenvalues

$r + \lambda_1(n - 1) + \lambda_2(mn - n)$ *(multiplicity 1),*

$r - \lambda_1$ *(multiplicity $m(n - 1)$),*

$r + \lambda_1(n - 1) - \lambda_2 n$ *(multiplicity $m - 1$).*

If the parameters in the matrix (1.5) satisfy (1.3), then the eigenvalues can be rewritten as rk, $r - \lambda_1$ and $rk - \lambda_2 mn$.

Proof. An eigenvector to the first eigenvalue is the all-one-vector. Let $a = (a_0, \ldots, a_{mn-1})^t$ be a vector of size mn. The vectors with $a_j = 1$ $(0 \le j \le m-1)$ and $a_{im+j} = -1$ $(1 \le i \le n-1)$ and $a_t = 0$ for the other coefficients are $mn - m$ linearly independent eigenvectors with eigenvalue $r - \lambda_1$. The eigenspace of $r + \lambda_1(n-1) - \lambda_2 n$ can be generated by the $m - 1$ vectors $a^{(j)} = (a_0^{(j)}, \ldots, a_{mn-1}^{(j)})$

$(1 \le j \le m - 1)$ with $a_{im}^{(j)} = 1$ and $a_{im+j}^{(j)} = -1$ for all i with $0 \le i \le n - 1$. $\square$

This motivates the "classification" of divisible designs due to Bose and Connor [32]. They call a divisible design **semiregular** if $r - \lambda_1 \ne 0$ but $rk - \lambda_2 mn = 0$. A design is **regular** if $r - \lambda_1 \ne 0$ and $rk - \lambda_2 mn \ne 0$. Note that the first eigenvalue is (with a very trivial exception) always different from 0. If $r = \lambda_1$, then the designs are also rather trivial: The blocks of the divisible designs are just unions of point classes. When we "contract" these point classes to new points, we get a design in the usual sense (see the definition below), and therefore the divisible designs with $r - \lambda_1$ are basically the same objects as designs in the usual sense. The term "regular" refers to the fact that the matrix in (1.5) is regular (over a field of characteristic 0). In the square case this shows that an incidence matrix of the design is regular. In general, we call a square design **regular** if an incidence matrix is regular over the rationals. Result 1.3.7 shows an interesting property of regular designs.

The case of divisible designs with $n = 1$ (or $\lambda_1 = \lambda_2 = \lambda$) is of particular interest. In this case, any two distinct points are joined by the same number λ of blocks, and we do not have to distinguish between different point classes. Then we call the design an (m, k, λ)-**design**: The design consists of m points such that any two distinct points are joined by exactly λ blocks. Moreover, the size of a block is k. A very important parameter of an (m, k, λ)-design is its order $r - \lambda$. If the design is square, then (1.4) shows $r = k$, and the order is simply $k - \lambda$. Since the order of a design is such a fundamental parameter, we will sometimes add it to the parameter set of the design and we speak about $(m, k, \lambda; n)$-designs. For instance, the projective plane of order 2 in Example 1.1.1 is a (square) $(7, 3, 1; 2)$-design. In general, a **projective plane** is a square design with $\lambda = 1$. If the order is denoted by n, then equation (1.3) (where we have to put $n = 1$ since we are in the design case) shows $m = n^2 + n + 1$, i.e. we can write the parameters of a projective plane in the form $(n^2 + n + 1, n + 1, 1)$. What can we say about the existence of such planes? We take the field $\mathrm{GF}(q)$ and define the 1-dimensional subspaces of the vector space $\mathrm{GF}(q)^{d+1}$ to be the points of an incidence structure. The blocks are the hyperplanes of this vector space. Then it is just an exercise from Linear Algebra to show that this incidence structure is a (square) design with parameters

$$\left(\frac{q^{d+1} - 1}{q - 1}, \frac{q^d - 1}{q - 1}, \frac{q^{d-1} - 1}{q - 1}; q^{d-1} \right).$$

These are the classical point-hyperplane designs of the projective geometry $\mathrm{PG}(d, q)$. (The projective geometry consists not just of points and blocks but contains the lattice of all subspaces of $\mathrm{PG}(d, q)$. If we take i-dimensional subspaces of $\mathrm{GF}(q)^{d+1}$ to be the blocks ($i \ne d$), then we obtain designs with more blocks than points.) The point-hyperplane designs of $\mathrm{PG}(d, q)$ are projective planes of order q if and only if $d = 2$. These projective planes are called **desarguesian** since they contain the so called *Desargues configuration*. There are other examples of projective planes (non-desarguesian), however, in all known examples the order is a prime power. This leads to one of the most striking open problems in finite geometry:

Problem 1 (Prime Power Conjecture) Do projective planes exist whose order is not a prime power?

We will investigate projective planes with a certain type of automorphism group in Chapter 5, where we will also say more about the prime power conjecture and related questions. The reader is referred to Hughes and Piper [91] or Pickert [139] for more on projective planes, in particular, for many constructions of non-desarguesian planes. Here we just want to mention that there is an equivalent definition of projective planes: When we say that planes are square designs we assume that the line size is constant. But it is enough to assume that the incidence structure contains a quadrangle. More precisely: Let $\mathcal{D}$ be an incidence structure, where any two distinct points are joined by exactly one line, and where any two distinct lines intersect. If this incidence structure contains a quadrangle, i.e., if there are four points no three of which are on a line, then the incidence structure is a projective plane. (Sometimes we will need this characterization). It is easy to prove that the existence of a quadrangle implies constant line size, see Hughes and Piper [91] or Pickert [139], for instance. If the incidence structure does not contain a quadrangle then there are "degenerate" examples.

The prime power conjecture would be solved if the parameters of all known square designs with $\lambda = 1$ were determined. A more difficult problem would be to determine the parameters of <u>all</u> possible (square) divisible designs. In this direction, only the following result is known. To state this result, we define the *Hilbert symbol* $(a, b)_p$ for integers a and b and primes p to be 1 or -1 depending on whether the congruence $ax_r^2 + by_r^2 \equiv z_r^2 \bmod p^r$ has or has not an integral solution for all r (where not all of the x_r, y_r, z_r should be 0).

Result 1.1.5 (Bose, Connor [32]) *Let $\mathcal{D}$ be a square regular divisible design with parameters $(m, n, k, \lambda_1, \lambda_2)$, i.e. $k^2 - mn\lambda_2 > 0$ and $k - \lambda_1 > 0$. Then the following holds:*

(a) *If m is even, then $k^2 - mn\lambda_2$ is a square. If moreover $m \equiv 2 \bmod 4$ and n is even, then $k - \lambda_1$ is the sum of two squares.*

(b) *If m is odd and n is even, then $k - \lambda_1$ is a square, and*

$$(k^2 - mn\lambda_2, (-1)^{(m-1)/2}n\lambda_2)_p = 1$$

for all odd primes.

(c) *If both m and n are odd, then*

$$(k - \lambda_1, (-1)^{(n-1)/2}n)_p \cdot (k^2 - mn\lambda_2, (-1)^{(m-1)/2}n\lambda_2)_p = 1$$

for all odd primes p.

Corollary 1.1.6 (Bruck-Ryser Theorem) *Let $\mathcal{D}$ be a square (m, k, λ)-design of order $k - \lambda$. If m is even, then $k - \lambda$ is a square. If m is odd, then*

$$(k - \lambda, (-1)^{(m-1)/2})_p = 1$$

for all odd primes p, equivalently the diophantine equation

$$(k - \lambda)x^2 + (-1)^{(m-1)/2}\lambda y^2 = z^2$$

has an integer solution where not all of x, y and z are 0.

Proof. We just put $n = 1$ in the Bose-Connor Theorem. For the equivalence between the statement about the Hilbert residue symbol and the solution of the diophantine equation, we refer the reader to Cassels [41], for instance. □

This Corollary has been proved by Schutzenberger [158] for the case that m is even. Bruck and Ryser [39] have considered the $\lambda = 1$ case (projective planes), the general version is in Chowla and Ryser [46].

The dual of a square (m, k, λ)-design is again an (m, k, λ)-design, therefore we can also speak about symmetric designs. We refer the reader to Beth, Jungnickel and Lenz [29] for a proof of this statement and also for a proof of Corollary 1.1.6.

Trivial examples of square (m, k, λ)-designs are defined by the incidence matrices $\mathbf{J}_m$ and $\mathbf{J}_m - \mathbf{I}_m$ with parameters (m, m, m) and $(m, m - 1, m - 2)$. The orders of these **trivial** designs are 0 and 1. Similarly, there are trivial examples of square divisible designs with parameters $(m, n, n, n, 0)$ (incidence matrix $\mathbf{J}_n \otimes \mathbf{I}_m$) and $(m, n, n - 1, n - 2, 0)$ (incidence matrix $(\mathbf{J}_n - \mathbf{I}_n) \otimes \mathbf{I}_m$). The complements of these designs are also called trivial: The **complement** of a square design $\mathcal{D}$ with incidence matrix $\mathbf{A}$ of size v is the design $\overline{\mathcal{D}}$ with incidence matrix $\mathbf{J}_v - \mathbf{A}$. The proof of the following Lemma is quite straightforward:

Lemma 1.1.7 *If $\mathcal{D}$ is a square divisible design with parameters $(m, n, k, \lambda_1, \lambda_2)$, then the complement $\overline{\mathcal{D}}$ is a square design with parameters $(m, n, mn - k, \lambda_1 + mn - 2k, \lambda_2 + mn - 2k)$.*

Proof. Let $\mathbf{A}$ be the incidence matrix of $\mathcal{D}$. We compute

$$(\mathbf{J}_{mn} - \mathbf{A})(\mathbf{J}_{mn} - \mathbf{A})^t = (mn - 2k)\mathbf{J}_{mn} + (k - \lambda_1)\mathbf{I}_{mn} + (\lambda_1 - \lambda_2)\mathbf{J}_n \otimes \mathbf{I}_m + \lambda_2 \mathbf{J}_{mn}$$

and compare the result with (1.2) (note $r = k$). □

This lemma shows that the complements of our trivial designs have parameters $(m, 0, 0)$, $(m, 1, 1)$, $(m, n, mn - n, mn - n, mn - 2n)$, and $(m, n, mn - n + 1, mn - n, mn - 2n + 1)$. Moreover, the parameters $k - \lambda_1$ and $\lambda_1 - \lambda_2$ of a divisible design are invariant under complementation, in particular, the "interesting" eigenvalues of $\mathbf{A}\mathbf{A}^t$ are invariant (the eigenvalue k^2 changes upon complementation).

In the investigation of square (m, k, λ)-designs, the designs are often replaced by their complements, which are designs of the same order. In case of divisible designs, we can do the same. However, sometimes we want to restrict ourselves to designs with $\lambda_1 = 0$, and in this case the complement of such a design has a λ_1-value different from 0. In other words, the class of divisible designs with $\lambda_1 = 0$ is not closed under complementation.

In the general case of "designs", we cannot think of blocks just as subsets of the point set. It is possible that two blocks consist of the same points but are different, for instance, if $\mathcal{D}$ is a (trivial) divisible $(m, n, n, n, 0)$-design or an (m, m, m)-design (there are less trivial examples!). However, it is easy to check that these are the only examples of square divisible designs where two different blocks consist of the same points (i.e., where an incidence matrix has two columns which are equal). A design is called **simple** if there are no two distinct blocks which consist of the the same points. In this case, we can write "$p \in B$" if the point p is incident with B.

We can formulate the problem of the design theorist to find 0/1 solutions for a matrix equation (1.1), where $\mathbf{M}$ is a suitable matrix as on the right hand side of (1.2). This seems to be a rather difficult problem. Under the assumption that the incidence structure $\mathcal{D}$ admits automorphisms, there are more algebraic techniques available to find solutions: An **automorphism** of a design $(\mathcal{V}, \mathcal{B}, I)$ is a bijection $\varphi : \mathcal{P} \cup \mathcal{B} \rightarrow \mathcal{P} \cup \mathcal{B}$ which maps points to points and blocks to blocks $(\varphi[\mathcal{P}] = \mathcal{P},\ \varphi[\mathcal{B}] = \mathcal{B})$, and which preserves incidence as well as non-incidence $((p, B) \in I \Leftrightarrow (\varphi(p), \varphi(B)) \in I)$. Automorphisms of designs are sometimes called **collineations**, in particular in the case of projective planes. Similarly, we define isomorphisms between designs and say that designs are isomorphic if there are isomorphisms between them. In this monograph, we will find several results where we assume that an automorphism group acts sharply transitively on points *and* blocks: A group G acts **sharply transitively** or **regularly** on a set X if for any two elements $x, y \in X$ there is exactly one group element $g \in G$ with $g(x) = y$. This implies $|G| = |X|$, and therefore it makes sense to speak about sharply transitive groups on the points and the blocks of a design only if this design $\mathcal{D}$ has the same number of points as blocks, i.e. if $\mathcal{D}$ is a square incidence structure. In this case, we have $v = b$, equivalently $r = k$.

An automorphism group G of a square design $\mathcal{D}$ that acts regularly on points and blocks is called a **Singer group** of $\mathcal{D}$. This is in honour of J. Singer who was the first author who investigated regular groups of the classical design formed by the points and hyperplanes of the projective geometry $\mathrm{PG}(d, q)$, see [161].

If a design admits a Singer group, there is a nice way to represent the design. We choose a base point p_0 and identify every point q with the (unique) group element which maps p_0 to q. We will think of blocks as point sets, hence blocks become subsets of the group G. (We do not encounter any difficulties if the design is one of the trivial non-simple designs.) We are now going to investigate some basic properties of these subsets. In order to do this, we choose group invariant incidence matrices which describe incidence structures with Singer groups: Let $\mathcal{D}$ admit a (multiplicatively written) Singer group G. Then we can choose an incidence matrix $\mathbf{A} = (a_{p,B})$ in such a way that $(a_{p,B}) = (a_{g(p),g(B)})$ for all $g \in G$. A matrix with this property is called a **group invariant** incidence matrix. The set of group invariant matrices (with group G) over a ring R forms itself a ring which is isomorphic to the group ring $R[G]$ (we assume that R is a commutative ring with identity 1). The **group ring** $R[G]$ is the set of formal sums

$$S = \sum_{g \in G} s_g g$$

with $s_g \in R$. Addition in $R[G]$ is defined componentwise

$$\sum_{g \in G} s_g g + \sum_{g \in G} t_g g = \sum_{g \in G} (s_g + t_g) g,$$

multiplication imitates the usual "convolution" of polynomials

$$\sum_{g \in G} s_g g \sum_{g \in G} t_g g = \sum_{h,k \in G} (s_h t_k) hk.$$

If R is actually a field, we call $R[G]$ a **group algebra**.

It turns out that the transpose of a group invariant matrix has a nice representation as an element in the group ring. Let us state these facts in the following proposition.

Proposition 1.1.8 *Let* $\mathcal{R}$ *denote the algebra of group invariant* $(v \times v)$-*matrices* $\mathbf{B}$ *over a ring* R, *i.e.,*

$$\mathbf{B} = (b_{i,j}) = (b_{g(i),g(j)})$$

for all $g \in G$ *where* G *acts regularly on* $\{1, \dots, v\}$. *Then*

$$\psi: \quad \begin{matrix} \mathcal{R} & \to & R[G] \\ (b_{i,j}) & \mapsto & \sum_{g \in G} b_{1,g(1)} g^{-1} \end{matrix}$$

is an algebra isomorphism and

$$\psi(\mathbf{B}^t) = \sum_{g \in G} b_{1,g(1)} g.$$

Proof. Note that the first row of a group invariant incidence matrix determines the entire matrix, therefore ψ is a bijection. Moreover,

$$\psi(\mathbf{AB}) = \sum_{g,h} (a_{1,g(1)} b_{g(1),h(1)}) h^{-1}$$

$$= \sum_h \left(\sum_g a_{1,g(1)} b_{1,g^{-1}(h(1))} \right) h^{-1}$$

and

$$\psi(\mathbf{A})\psi(\mathbf{B}) = \sum_{g,k} (a_{1,g(1)} b_{1,k(1)}) g^{-1} k^{-1}$$

$$= \sum_h \left(\sum_g a_{1,g(1)} b_{1,g^{-1}(h(1))} \right) h^{-1} \quad \text{(put } h = kg\text{)}.$$

The assertion about $\psi(\mathbf{B}^t)$ is obvious.   □

By abuse of notation, we will call the group ring element $\psi(\mathbf{A})$ that can be associated with each group invariant matrix simply A. If $A = \sum_{g \in G} a_g g$, we define

$$A^{(s)} := \sum_{g \in G} a_g g^s$$

for every integer s. In general, $A^{(s)}$ is completely different from A^s, but we have

$$A^{(p)} = A^p \quad \text{in } K[G]$$

if K is a field of characteristic p.

Now we can translate the matrix equation (1.1) into an equation

$$AA^{(-1)} = M \qquad\qquad (1.6)$$

in $R[G]$. If $\mathbf{A}$ is a (group invariant) incidence matrix, then $\mathbf{A}$ has $0/1$-coefficients, and the same is true for the group ring element $\psi(\mathbf{A}) = A$. Note that a group ring element with coefficients 0 and 1 naturally describes a subset of G (it is the "characteristic vector" of a subset of G). How do we obtain this subset from the group invariant incidence matrix $\mathbf{A}$? Let p_0 (resp. B_0) be the point (resp. block) corresponding to the first row (resp. column) of $\mathbf{A}$. Then we have $g^{-1} \in A$ if and only if $a_{1,g} = 1$, which is true if and only if $g^{-1}(p_0) \in B_0$. In other words, A consist of the elements which map the point corresponding to the first row of a group invariant matrix $\mathbf{A}$ to a point which is on the block corresponding to the first column of $\mathbf{A}$. The point p_0 is called the **base point**, and the block B_0 is the **base block**. Conversely, if we choose a base point and a base block arbitrarily, we can construct a group invariant matrix where the first row (column) corresponds to the base point (block). Thereforeilizer the set A corresponding to a group invariant incidence matrix "is" a block of the design (if the points are identified with the group elements). The translates $Ag := \{ag : a \in A\}$ are all b $(= v)$ blocks of $\mathcal{D}$!

We will illustrate this in the case of an $(m, n, k, \lambda_1, \lambda_2)$-divisible design $\mathcal{D}$. Let G be an automorphism group acting sharply transitively on the points and the blocks of $\mathcal{D}$. We choose a point p_0 to be our base point. This point is contained in a point class, and we denote the stabilizer of this point class by N, which is a subgroup of G of order n. The point classes are the (right) cosets of N if we identify points with group elements: Two points $g(p_0)$ and $h(p_0)$ are in the same point class if and only if $h^{-1}(g(p_0))$ and p_0 are in the same point class, i.e. $gh^{-1} \in N$. Thus the group ring equation corresponding to a square divisible design with a Singer group becomes

$$AA^{(-1)} = (k - \lambda_1) + (\lambda_1 - \lambda_2)N + \lambda_2 G \qquad\qquad (1.7)$$

in $\mathbf{Z}[G]$: The "matrix" on the right hand side of (1.7) tells us that any two points $g(p_0)$ and $h(p_0)$ with $gh^{-1} \notin N$ are joined by exactly λ_2 blocks and that points with $gh^{-1} \in N$ $(g \neq h)$ are joined by λ_1 blocks.

Conversely, a group ring element A in $\mathbf{Z}[G]$ with coefficients 0 and 1 which satisfies (1.7) corresponds to a divisible design with a Singer group: We can partition the set of points into m point classes (which are the cosets of N) such that the number of blocks joining two different points depends on whether the points are in the same point class or not. We must be a little bit careful: The matrix on the right hand side of (1.6) or (1.7) is in general not the matrix $\mathbf{M}$ on the right hand side of (1.1) or (1.2). But in order to get a divisible design, it is enough that the matrix corresponding to the right hand side of (1.7) can be transformed via row and column permutations to the matrix in (1.2), and this is always possible.

Example 1.1.9 The matrix

$$
\left(
\begin{array}{cccc|cccc}
1 & 1 & 0 & 0 & 1 & 0 & 0 & 1 \\
0 & 1 & 1 & 0 & 1 & 1 & 0 & 0 \\
0 & 0 & 1 & 1 & 0 & 1 & 1 & 0 \\
1 & 0 & 0 & 1 & 0 & 0 & 1 & 1 \\
\hline
1 & 0 & 0 & 1 & 1 & 1 & 0 & 0 \\
1 & 1 & 0 & 0 & 0 & 1 & 1 & 0 \\
0 & 1 & 1 & 0 & 0 & 0 & 1 & 1 \\
0 & 0 & 1 & 1 & 1 & 0 & 0 & 1
\end{array}
\right)
$$

is (canonically) invariant under the group $\mathbf{Z}_4 \times \mathbf{Z}_2$. If we compute

$$
\mathbf{A}\mathbf{A}^t =
\left(
\begin{array}{cccccccc}
4 & 2 & 0 & 2 & 2 & 2 & 2 & 2 \\
2 & 4 & 2 & 0 & 2 & 2 & 2 & 2 \\
0 & 2 & 4 & 2 & 2 & 2 & 2 & 2 \\
2 & 0 & 2 & 4 & 2 & 2 & 2 & 2 \\
2 & 2 & 2 & 2 & 4 & 2 & 0 & 2 \\
2 & 2 & 2 & 2 & 2 & 4 & 2 & 0 \\
2 & 2 & 2 & 2 & 0 & 2 & 4 & 2 \\
2 & 2 & 2 & 2 & 2 & 0 & 2 & 4
\end{array}
\right)
$$

we see that the product $\mathbf{A}\mathbf{A}^t$ is not the matrix $\mathbf{M}$ on the right hand side of (1.2). But it is easy to transform this matrix via row and column permutations to such an $\mathbf{M}$.

The group ring equation (1.7) can be reformulated as follows: The group ring element A "is" a k-subset of a group G of order mn such that every element outside a subgroup N of order n has exactly λ_2 representations as a quotient dd'^{-1} with elements $d, d' \in A$. Elements in N different from the identity have exactly λ_1 such representations. (In the example above, this set is $\{(0,0),(3,0),(0,1),(1,1)\} \subset \mathbf{Z}_4 \times \mathbf{Z}_2$.) Any set with this property is called an $(m, n, k, \lambda_1, \lambda_2)$-**divisible difference set** in G relative to N. If $\lambda_1 = 0$, we call it a **relative difference set**. In this case, we call the exceptional subgroup N the **forbidden subgroup**. If $n = 1$ (in which case we do not need a λ_1-value and simply write λ instead of λ_2), we speak about (m, k, λ)-**difference sets**.

In this case, we call $k - \lambda$ the **order** of the difference set, which is the same as the order of the (m, k, λ)-design associated with the difference set. Difference sets are called *abelian, cyclic* etc., if the group that contains the difference set has this property. A divisible difference set in G is called **splitting** if the exceptional subgroup N is a direct factor of G, i.e. N has a complement in G. We will now give some small examples of (divisible) difference sets.

Example 1.1.10   1. $D = \{0, 1, 3\} \subset \mathbf{Z}_7$ is a cyclic $(7, 3, 1)$-difference set.

2. $R = \{0, 1, 3\} \subset \mathbf{Z}_8$ is a cyclic relative $(4, 2, 3, 1)$-difference set.

3. This is a non-cyclic example: The set $\{(0, 0), (1, 1), (2, 1)\} \subset \mathbf{Z}_3 \times \mathbf{Z}_3$ is a relative $(3, 3, 3, 1)$-difference set.

4. The set $\{1, i, j, k\}$ in the quaternion group is a relative $(4, 2, 4, 2)$-difference set relative to the unique subgroup $\{\pm 1\}$ of order 2 in the quaternion group.

5. The set $\{1, 2, 4, 8\}$ in $\mathbf{Z}_{12}$ is a cyclic divisible difference set with parameters $(6, 2, 4, 2, 1)$.

6. Let $G \cong \mathbf{Z}_{13} \times S_3$ (where S_n denotes the symmetric group on n letters). Then the set

$$\{(1, id), (10, id), (11, id), (0, (12)), (5, (12)),$$
$$(2, (02)), (8, (02)), (7, (01)), (9, (01))\}$$

is a $(13, 6, 9, 1)$-difference set relative to S_3.

7. We are now going to describe an example of a difference set relative to a subgroup which is not normal. Let $G \cong \langle x, y : x^{13} = y^6 = 1, y^{-1}xy = x^4 \rangle$. The set

$$\{x^0 y^0, x^1 y^0, x^5 y^3, x^5 y^4, x^7 y^0, x^7 y^0, x^9 y^1, x^9 y^5, x^{10} y^2\}$$

is a $(13, 6, 9, 1)$-difference set relative to $\langle y \rangle$.

   In view of the last example in 1.1.10, some remarks about the normality of the subgroup N are in order. In the discussion so far it is not necessary that N is a normal subgroup: The subgroup N is the stabilizer of the point class through the base point p_0, and N acts regularly on this point class. If we identify points with group elements, then the (right) cosets of N are the point classes. The stabilizer of the point class is normal in G if it stabilizes all the point classes (since G acts transitively). Then we say that the Singer group is **normal**. This condition is satisfied in most of the known examples. In the past, most people considered abelian examples (and in the preface I have tried to explain why). It is possible that we will find many examples of non-normal Singer groups as soon as we try to find more non-abelian examples. But it is also possible that the non-normal case is the exception, I simply do not know. Let us summarize the discussion so far in a theorem:

Theorem 1.1.11 *The existence of a divisible design $\mathcal{D}$ with parameters $(m, n, k, \lambda_1, \lambda_2)$ and Singer group G is equivalent to the existence of a divisible $(m, n, k, \lambda_1, \lambda_2)$-difference set R in G relative to N. The subgroup N is the (setwise) stabilizer of some point class of $\mathcal{D}$. The incidence structure formed by the elements of G as points and the translates Rg as blocks is isomorphic to $\mathcal{D}$. Moreover, N is a normal subgroup if and only if N is the stabilizer of all point classes.* □

Quite often we denote divisible and relative difference sets by "R" and the "usual" difference sets by D. In the literature, the term diference set is widely used and established. In more theoretical investigations people usually write the group multiplicatively and nevertheless speak about difference sets. On the other hand, most (at least the classical) examples can be found in abelian groups. In this case people prefer additive notation. In this monograph, I will use both the additive and multiplicative notion for groups: The more or less theoretical the investigation the more or less likely is it that the groups are written multiplicatively.

The existence of a <u>relative</u> difference set implies the existence of a "series" of difference sets via a projection procedure. In general, let R be a divisible difference set in G relative to N. Let U be a normal subgroup of G. We extend the canonical epimorphism $\rho : G \rightarrow G/U$ to an epimorphism from $\mathbf{Z}[G] \rightarrow \mathbf{Z}[G/U]$. Then $\rho(D)$ satisfies the equation

$$\rho(D)\rho(D)^{(-1)} = (k - \lambda_1) + (\lambda_1 - \lambda_2)|U \cap N|\rho(N) + \lambda_2|U|\rho(G) \qquad (1.8)$$

in $\mathbf{Z}[G/U]$.

We must warn the reader: By $\rho(G)$ or $\rho(N)$ we denote subgroups of G/U and not the image of the group ring element G under ρ: If we consider G (or N or any other subgroup) as an element in $\mathbf{Z}[G]$, then $\rho(G)$ would be a group ring element with coefficients $|U|$. On the other hand, by $\rho(D)$ we denote the image of the group ring element D. This is the reason why several people use a different notation for subsets D of G and the corresponding group ring elements $\overline{D}$. In this mongraph, we do not have a special notation for group ring elements: It should always be easy to see whether we apply an epimorphism ρ to a subset of G or the corresponding group ring element. As a general rule, the image $\rho(T)$ denotes a subgroup in G/U if T is a subgroup of G (and where ρ is the canonical epimorphism from G onto G/U).

Since the group G/U is smaller than the group G, it is sometimes easier to show that (1.8) has no solution than to show this for (1.7). But the problem is, of course, that the element $\rho(D)$ has not just 0/1-coefficients, but now the coefficients are integers in the interval $[0, |U|]$, which makes life more difficult. In the case of a <u>relative</u> difference set we can improve the bound on the parameters of $\rho(D)$: Since every coset of N contains at most one element of R, each coset of U contains at most $|U|/|U \cap N|$ elements. In particular, if $U \subset N$, the projection of a relative difference set is a relative difference set again, and if $U = N$, this relative difference set is a difference set in the usual sense. Let us summarize this in the following Lemma:

Lemma 1.1.12 (Elliott, Butson [68]) *Let R be an (m, n, k, λ)-difference set in G relative to N. If U is a normal subgroup of G contained in N, and if ρ denotes the canonical epimorphism $G \to G/U$, then $\rho(R)$ is an $(m, n/u, k, \lambda u)$-difference set in G/U relative to N/U. In particular, there exists an $(m, k, n\lambda)$-difference set in G/N.*   □

This Lemma shows that we can think of relative difference sets as "extensions" or "liftings" of difference sets. Relative difference sets are not only a generalization of difference sets in the sense that each difference set is also a relative difference set, but the existence of a relative difference set <u>implies</u> the existence of a difference set.

The projection of $\mathbf{Z}[G]$ onto $\mathbf{Z}[G/U]$ is of course not injective (unless $|U| = 1$). All we can say is the following:

Lemma 1.1.13 *Suppose that $A = \sum a_g g$ and $B = \sum b_g g$ are elements in $R[G]$ and let U be a normal subgroup of G. Let ψ denote the canonical epimorphism from G onto G/U and also its extension to the group rings. If $\psi(A) = \psi(B)$, then $UA = UB$.*

Proof. Note that $\sum_{g \in Uh} a_g$ is the coefficient of $\psi(h)$ in $\psi(A)$. But it is also the coefficient of h in UA. The same holds for B which proves the Lemma.   □

We say that two difference sets are **isomorphic** if the corresponding designs are isomorphic. Sometimes difference sets are called isomorphic only if they are defined in isomorphic groups. Examples are known that difference sets in non-isomorphic groups describe isomorphic designs: The desarguesian projective planes of order $n \equiv 1 \bmod 3$ admit both a cyclic and a Frobenius group as a Singer group. This result is due to Bruck [38], see also Gao and Wei [74] and Pott [146] for generalizations. Another (stronger) relation is the equivalence of difference sets: We say that two difference sets D and F in G are **equivalent** if there is a group automorphism that maps D onto some translate Fg, $g \in G$. Note that the translates Fg are the blocks of the incidence structure described by F. Thus equivalent difference sets are isomorphic! It should be noted that sometimes a different notion of equivalence is used: It is only required that D can be mapped onto hFg for suitable elements $h, g \in G$. Of course, in the case of abelian difference sets both definitions coincide. Moreover, we note that isomorphism of two difference sets (if they are defined in isomorphic groups) does not imply equivalence of the difference sets, see Kibler [108]. However, no **cyclic** example seems to be known where this phenomenon occurs.

Problem 2 Try to find examples of cyclic difference sets which are isomorphic but not equivalent.

Divisible designs are a very special type of "partially balanced incomplete block designs (PBIBD)", see Chapter 8 in Raghavarao [153], for instance. In that more general case, we have d associate classes: Two distinct points are 1st, 2nd,...,d-th associate and the number of blocks joining them depends on which association class they share. The association classes have to form a symmetric association scheme, see Bannai and Ito [25], for instance. The defining matrix equation for PBIBD's is

$$\mathbf{A}\mathbf{A}^t = k\mathbf{I} + \sum_{i=1}^{d} \lambda_i \mathbf{N}_i,$$

where the N_i's are the association matrices, and $\mathbf{I}$ is the identity matrix. Again, if a square PBIBD admits a Singer group G, this equation can be translated into a group ring equation. Thus a PBIBD with Singer group corresponds to some subset D of G whose list of quotients different from 1 takes just d distinct values depending on the underlying association scheme. In the case of divisible designs, this association scheme has just two classes (thus it is a so called *strongly regular graph*), namely, the disjoint union of m copies of the complete graph on n vertices. Recently, Ma [126] has investigated the existence question for PBIBD's with Singer groups where the underlying association scheme is not such a trivial scheme.

We will finish this introductory section with some remarks concerning the question whether PBIBD's with Singer groups are symmetric. The association matrices $\mathbf{N}_i$ correspond to subgroups N_i (by abuse of notation) of the Singer group G: The group N_i is just the stabilizer of the set of points which are i-th associate to the base point p_0. If the subgroups N_i are <u>normal</u> subgroups, then the group ring element $M := k + \sum_{i=1}^{d} \lambda_i N_i$ in $\mathbf{Q}[G]$ is in the center of $\mathbf{Q}[G]$. Let D be the difference set which describes the PBIBD. If M is invertible, then D is invertible (in $\mathbf{Q}[G]$) and we get

$$DD^{(-1)} = M,$$
$$D^{(-1)} = D^{-1}DD^{(-1)} = D^{-1}M = MD^{-1},$$
$$D^{(-1)}D = M,$$

therefore, an incidence matrix $\mathbf{D}$ of the PBIBD satisfies

$$\mathbf{D}\mathbf{D}^t = \mathbf{D}^t\mathbf{D}$$

and the design is symmetric. In the case of square divisible designs, the right hand side of (1.2) and hence of (1.7) is invertible if and only if k, $k - \lambda_1$ and $k^2 - \lambda_2 mn$ are non-zero since these are the eigenvalues of the matrix on the right hand side of (1.2) If we are in the square case, this implies that the incidence matrices on the left hand side of (1.2) are regular, too. Jungnickel [95] proved that any square divisible design with a normal Singer group is symmetric provided that not both $\lambda_1 \neq 0$ and $k^2 = mn\lambda_2$ holds. In his paper, Jungnickel erroneously claimed that difference sets with these parameters cannot exist. However, examples of such difference sets are known, see Section 2.3.

1.2   Tools from representation theory and algebraic number theory

Powerful methods to find necessary conditions on the existence of difference sets come from algebraic number theory as well as representation theory. We will begin with the representation theoretic part. The difference set will be interpreted as an element in the group algebra $K[G]$ where K is some field. Let us assume that the characteristic of the field and the group order are relatively prime. Then it is well known that the group algebra is semi-simple, which means that the algebra can be decomposed as an algebra into simple algebras (which are algebras without non-trivial two-sided ideals). The reader is refered to any good text book on algebra for proofs of the facts that we will use throughout this monograph (for instance, Curtis and Reiner [48] or Huppert [92]) We will assume that G is an <u>abelian</u> group of order v. Then there are exactly v distinct 1-dimensional representations of G over an extension field E of K: In order to "see" all these representations, we have to assume that E contains a v^*-th root of unity where v^* is the **exponent** of G (which is the smallest non-negative integer s satisfying $g^s = g$ for all $g \in G$). The 1-dimensional representations are called **characters**, and they are nothing else than homomorphisms from G into the multiplicative group of E. The characters form a group called the **character group** G^*, which is isomorphic with G. The identity element of G^* is called the **principal** character χ_0: It is the homomorphism which maps every element to 1. In order to see all this, we decompose the group G into a direct product of cyclic subgroups $C_i \cong \langle g_i \rangle$ with $|C_i| = w_i$. The characters are the mappings

$$\begin{aligned} \chi : \quad G &\rightarrow E \\ g_i &\mapsto \zeta \end{aligned}$$

where ζ is a w_i'th root of unity. This is the reason why we have to assume that E contains a primitive v^*-th root of unity.

Characters are defined as mappings from G into E. This mapping can be extended by linearity to an algebra homomorphism from $E[G]$ into E (consider E as an algebra on E). In a finite abelian group, the element $\chi(G)$ depends only on whether χ is the principal character or not. This is part of the fundamental orthogonality relations for characters of abelian groups:

Lemma 1.2.1 (orthogonality relations) *Let G be an abelian group of order v and exponent v^*. If the field E contains a primitive v^*-th root of unity (in particular, the characteristic of E and v are relatively prime), then*

$$\chi(G) = \begin{cases} |G| & \text{if } \chi = \chi_0 \\ 0 & \text{if } \chi \neq \chi_0 \end{cases} \tag{1.9}$$

for all characters of G. Moreover, we have

$$\sum_{\chi \in G^*} \chi(g) = \begin{cases} |G| & \text{if } g = 1 \\ 0 & \text{if } g \neq 1. \end{cases} \tag{1.10}$$

Proof. Let χ be a non-principal character. Then there is a group element $h \in G$ such that $\chi(h) \neq 1$. We get $\chi(G) = \chi(Gh) = \chi(G)\chi(h)$ which shows $\chi(G) = 0$. The remaining assertions follow in the same way. $\square$

There is a nice and powerful "duality" theory about characters. Let U be a subgroup of G. Then the set

$$\{\chi \in G^* : \chi(g) = 1 \text{ for all } g \in U\} =: U^\perp$$

is a subgroup of G^* of order G/U. The characters in $U^\perp$ are basically the characters of G/U: If $\chi' \in (G/U)^*$, then the mapping

$$\begin{array}{rcl} \chi : & G & \to & E \\ & g & \mapsto & \chi'(gU) \end{array}$$

is a character in $U^\perp$ and, conversely, each character in $U^\perp$ gives rise to a character of G/U. The duality theory is important for us to investigate divisible difference sets: In that case, we have an exceptional subgroup N in G. Corresponding to this subgroup, we have an exceptional subgroup N^* in the character group G^*, see (1.11).

Let us look at the cyclic case a little bit more carefully: Let H be a cyclic group of order p^t and let P_i denote the unique subgroup of order p^i ($i = 0, \ldots, t$). A character of order at most p^s is principal on P_{t-s}, i.e., the characters whose order divides p^s are contained in $P_{t-s}^\perp$. Conversely, the order of a character in $P_{t-s}^\perp$ divides p^s, hence

$$\{\chi \in H^* : \text{the order of } \chi \text{ divides } p^s\} = P_{t-s}^\perp$$

or

$$P_s^* = P_{t-s}^\perp$$

(where P_i^* denotes the unique subgroup of order p^i in H^*). In the general (non-cyclic) case, we have $U^\perp \cong (G/U)^*$, but there is not such a "canonical" way to identify $U^\perp$ as a subgroup of G^*.

If G is non-abelian, then there are precisely $|G/G'|$ distinct 1-dimensional characters of G (here G' denotes the commutator subgroup of G). This can be easily seen since G/G' is abelian, and every character χ' of G/G' can be "lifted" to a character of G via $\chi(g) := \chi'(gG')$.

Characters are very useful in connection with the next lemma called the "inversion formula". Roughly speaking, the character values uniquely determine the group algebra element:

Lemma 1.2.2 (inversion formula) *Let G be an abelian group of exponent v^* and let $A = \sum_{g \in G} a_g g$ be an element in the group algebra $E[G]$ where E contains a primitive v^*-th root of unity. Then*

$$a_g = \frac{1}{|G|} \sum_{\chi \in G^*} \chi(A)\chi(g^{-1})$$

where G^* is the character group of G.

Proof. We compute

$$\sum_{\chi \in G^*} \chi(Ag^{-1}) = \sum_{\chi \in G^*} \sum_{h \in G} a_h \chi(h)\chi(g^{-1}) = a_g|G|$$

using (1.10).   □

If A is an element in the integral group ring $\mathbf{Z}[G]$, then the complex character values $\chi(A)$ are not arbitrary complex numbers but algebraic integers. If the character χ has order w, then $\chi(A)$ is an element in $\mathbf{Z}[\zeta_w]$ where ζ_w is a primitive complex w-th root of unity. This is the reason why algebraic number theory is an important tool in order to investigate difference sets. We are now going to summarize those results from algebraic number theory which are used quite frequently. The reader can find proofs of these results in any good textbook about algebraic number theory, for instance Ireland and Rosen [93] or Weiss [172].

The first of our results shows that the ring $\mathbf{Z}[\zeta_w]$ is the same as the ring of algebraic integers of $\mathbf{Q}(\zeta_w)$. The second statement has many applications to difference sets, in particular, if $w = p$. By $\phi(d)$ we denote the Euler ϕ-function, i.e., the number of integers between 1 and $d-1$ which are relatively prime to d. Recall that an integral basis S is a set of algebraic integers in a field E such that every algebraic integer in E can be written uniquely as an integral linear combination with elements from S.

Result 1.2.3 *Let ζ_w be a primitive w-th root of unity in $\mathbf{C}$. Then $\mathbf{Z}[\zeta_w]$ is the ring of algebraic integers in $\mathbf{Q}(\zeta_w)$. Any set of $\phi(w)$ consecutive powers of ζ_w forms an integral basis.*

The next result will be needed quite often. Most of the applications of algebraic number theory to difference sets are related to it.

Result 1.2.4 *The algebraic integers of $\mathbf{Q}(\zeta_w)$ form a Dedekind domain, i.e. each ideal can be factored uniquely into prime ideals. Furthermore, two integers relatively prime in $\mathbf{Z}$ remain relatively prime in the ring of algebraic integers, in particular $\mathbf{Z}[\zeta] \cap \mathbf{Q} = \mathbf{Z}$.*

Using this result and the inversion formula (Lemma 1.2.2), we get the following important corollary:

Corollary 1.2.5 *Let $A = \sum a_g g$ be an element in $\mathbf{Z}[G]$ where G is an abelian group of exponent v^*. Let U be a subgroup of G. Suppose that $a_g \geq 0$ for all g. Let ζ be a primitive complex v^*-th root of unity, and let m be an integer with $(m, v^*) = 1$. If $\chi(A) \equiv 0 \bmod m$ for each character $G \rightarrow \mathbf{Q}(\zeta)$ (more precisely, its extension to the group algebra $\mathbf{Q}[G]$) which is non-principal on U, then $a_g \equiv a_{gu} \bmod m$ for all $u \in U$. In other words: The coefficients of A are constant modulo m on cosets of U. If $\chi(A) = 0$ for all characters non-principal on U, then $A = UX$ for some X in $\mathbf{Z}[G]$.*

Proof. The inversion formula shows

$$v(a_g - a_{gu}) = \sum_{\chi \in G^*} \chi(A)(\chi(g^{-1}) - \chi(g^{-1}u^{-1}))$$

$$= \sum_{\chi|U \neq \chi_0} \chi(A)(\chi(g^{-1}) - \chi(g^{-1}u^{-1})).$$

The right hand side of this equation is divisible by m in $\mathbf{Z}[\zeta]$ since this is true for each of the character values $\chi(A)$. Therefore, $v(a_g - a_{gu})/m$ is an element in $\mathbf{Z}[\zeta] \cap \mathbf{Q} = \mathbf{Z}$. But this is only possible if $a_g - a_{gu}$ is divisible by m.

The assertion for the case $\chi(A) = 0$ is analogous. $\qquad\square$

The following result strengthens this corollary. Its proof uses the integral basis mentioned in Result 1.2.3. We refer the reader to the original paper or Lander [116, Section 4.6]

Result 1.2.6 (Turyn [167]) *Let $G \cong K \times H$ be an abelian group, where K is cyclic of order v_1 and H is of order v_2. Suppose that $A = \sum a_g g$ is an element in $\mathbf{Z}[G]$ with $a_g \geq 0$. Let H^* be the character group of H and χ_1 be a generator of the character group K^* of K. We assume $v_1 \neq 1$ and $m|\chi_1\chi(A)$ for all characters $\chi \in H^*$ and some integer m relatively prime to v_2 (note $G^* \cong K^* \times H^*$). We assume $v_1 \neq 1$ and $\chi_1\gamma(A) \neq 0$ for at least one character $\gamma \in H^*$. Then*

$$m \leq 2^{r-1} max\{a_g\},$$

where r denotes the number of distinct prime divisors of v_1.

The decomposition of a prime p into prime ideals in $\mathbf{Z}[\zeta_w]$ is explicitely known. We have the following result:

Result 1.2.7 *Let p be a prime and ζ_w a primitive w-th root of unity in $\mathbf{C}$.*

(a) *If $w = p^e$, then the decomposition of the ideal (p) in $\mathbf{Q}(\zeta_w)$ into prime ideals is $(p) = (1 - \zeta_w)^{\phi(w)}$.*

(b) *If $(w, p) = 1$, then the prime ideal decomposition of (p) is $(p) = \pi_1 \ldots \pi_g$, where the π_i's are distinct prime ideals. Furthermore, $g = \phi(w)/f$ where f is the order of p modulo w. The field automorphism $\zeta_w \to \zeta_w{}^p$ fixes the ideals π_i.*

(c) *If $w = p^e w'$ with $(w', p) = 1$, then the prime ideal (p) decomposes as $(p) = (\pi_1 \ldots \pi_g)^{\phi(p^e)}$, where the π_i's are distinct prime ideals and $g = \phi(w')/f$. Here f is defined as in (b). If t is an integer not divisible by p and $t \equiv p^s \bmod w'$ for a suitable integer s, then the field automorphism $\zeta_w \to \zeta_w{}^t$ fixes the ideals π_i.*

Also the following results are useful. The first is sometimes attributed to Kronecker.

Result 1.2.8 (Kronecker) *Let η be an algebraic integer in $\mathbf{Q}(\zeta)$ where ζ is some root of unity. If η and all its algebraic conjugates have absolute value at most 1, then η is a root of unity.*

Result 1.2.9 (see Ireland, Rosen [93, Chapter 6]) *Let p be a prime and $A \in \mathbf{Z}[H]$ be an element in the integral group ring over the cyclic group $H = \langle h \rangle$ of order p. Then $\chi(A)\overline{\chi(A)} = p$ for all complex characters $\chi \neq \chi_0$ if and only if there exists a suitable translate Ag of A with*

$$Ag = xH + \sum_{i=0}^{p-1} \left(\frac{i}{p}\right) h^i$$

for some integer x. The integer x can be determined from the principal character value $\chi_0(A)$ (we have $\chi_0(A) = xp$).

(Here $\left(\frac{i}{p}\right)$ is the so called **Legendre symbol**: It is 0, 1 or -1 depending on whether i is 0, a square or a non-square modulo p.)

How can we apply all these results to difference set problems? We consider a group ring equation $AA^{(-1)} = M$ in $\mathbf{Z}[G]$. Let χ be a complex character. We get

$$\chi(A)\chi(A^{(-1)}) = \chi(M)$$

where $\chi(A^{(-1)}) = \overline{\chi(A)}$ is the complex conjugate of $\chi(A)$. In many cases we know $\chi(M)$. For instance, in case of (m, k, λ)-difference sets we have $M = n + \lambda G$ with $n = k - \lambda$. Then (1.9) shows $\chi(M) = n$ for non-principal characters. In the more general case of divisible difference sets, we have (using (1.3))

$$\chi(k - \lambda_1 + (\lambda_1 - \lambda_2)N + \lambda_2 G) = \begin{cases} k^2 & \text{if } \chi = \chi_0 \\ k^2 - \lambda_2 mn & \text{if } \chi \in N^\perp, \chi \neq \chi_0 \quad (1.11) \\ k - \lambda_1 & \text{if } \chi \notin N^\perp. \end{cases}$$

This poses the problem how to get information about $\chi(A)$ if the square $\chi(A)\overline{\chi(A)}$ of the absolute value of $\chi(A)$ is known. We begin with the case that the group G is a cyclic p-group of order p^t.

Lemma 1.2.10 *Let p be a prime and $H = \langle h \rangle$ be a cyclic group of order p^t. Suppose A is an element in $\mathbf{Z}[H]$ such that $\chi(A)\chi(A^{(-1)}) = p^s$ for every character χ of H of order p^t. Let*

$$X = \begin{cases} 1 & \text{if } s = 0 \\ p^{(s/2)-1}(p - \sum_{i=0}^{p-1} h^{i \cdot p^{t-1}}) & \text{if } s \geq 1 \text{ and } s \text{ is even} \\ p^{(s-1)/2} \sum_{i=0}^{p-1} \left(\frac{i}{p}\right) h^{i \cdot p^{t-1}} & \text{if } s \text{ is odd and } p \text{ is odd} \\ 2^{(s-3)/2}(1 + h^{2^{t-2}} - h^{2 \cdot 2^{t-2}} - h^{3 \cdot 2^{t-2}}) & \text{if } s \text{ is odd}, s \geq 3, p = 2, \\ & \qquad \text{and } t \geq 2 \\ 1 + h^{2^{t-2}} & \text{if } s = 1, p = 2 \text{ and } t \geq 2. \end{cases}$$

Then

$$A = \pm Xg + P_1 Y \quad in \ \mathbf{Z}[H]$$

for some $g \in H$ and $Y \in \mathbf{Z}[H]$ where P_1 is the unique subgroup of order p in H.

Proof. Let ζ be a primitive p^t-th root of unity and let χ be the character of order p^t on H which maps h to ζ. Then χ is non-principal on P_1 and, using Result 1.2.9, we get

$$\chi(X)\chi(X^{(-1)}) = |\chi(X)|^2 = p^s$$

and therefore (as an equation for ideals)

$$(\chi(X)) = (1 - \zeta)^{s \cdot \phi(p^t)/2} = (\chi(A)),$$

see Result 1.2.7. This shows $\chi(A) = u\chi(X)$ for some unit u in $\mathbf{Q}[\zeta]$. Note that $|u|^2 = 1$ because of our assumption on A. Any algebraic conjugate of u is of the form $\sigma(u)$ where $\sigma \in \mathrm{Gal}(\mathbf{Q}(\zeta)/\mathbf{Q})$ maps ζ to ζ^t for some t relatively prime to p. Since the composition of σ and χ is also a character of H of order p^t, we get $|\sigma(u)^2| = 1$, again. Then Kronecker's Theorem (Result 1.2.8) implies $u = \pm\zeta^i$, i.e. $\chi(A) = \pm\zeta^i\chi(X)$. This shows $\chi(A \mp g^i X) = 0$ for all characters non-principal on P_1 (which are precisely the characters of order p^t). Then Corollary 1.2.5 shows $A \mp g^i X = P_1 Y$. □

The case $p = 2$, s odd, and $t = 1$ cannot occur since there had to exist elements a and b in $\mathbf{Z}$ such that $(\chi(a + bh))^2 = (a - b)^2 = 2^s$ with s odd, which is impossible.

We can use Lemma 1.2.10 inductively.

Proposition 1.2.11 (Ma, Pott [127]) *Let p be a prime and $H = \langle h \rangle$ be a cyclic group of order p^t. Suppose A is an element in $\mathbf{Z}[H]$ such that $\chi(A)\chi(A^{(-1)}) = p^s$ for every character χ of H of order <u>at least</u> p^{t-n+1}, where $1 \leq n \leq t$ and $n \leq \lfloor \frac{s}{2} \rfloor + 1$. Let*

$$X_m = \begin{cases} 1 & \text{if } s \text{ is even and } m = s/2 \\ p^{(s/2)-m-1}(p - \sum_{i=0}^{p-1} h^{i \cdot p^{t-m-1}}) & \text{if } s \text{ is even and } m < s/2 \\ p^{(s-1)/2-m} \sum_{i=0}^{p-1} \left(\frac{i}{p}\right) h^{i \cdot p^{t-m-1}} & \text{if } s \text{ is odd and } p \text{ is odd} \\ 2^{(s-3)/2-m}(1 + h^{2^{t-m-2}} & \text{if } s \text{ is odd, } m \leq (s-3)/2 \\ \quad - h^{2 \cdot 2^{t-m-2}} - h^{3 \cdot 2^{t-m-2}}) & \quad \text{and } p = 2 \\ 1 + h^{2^{t-m-2}} & \text{if } s \text{ is odd, } p = 2, m = (s-1)/2 \\ & \quad \text{and } t \geq m + 2. \end{cases}$$

Then

$$A = \sum_{m=0}^{n-1} \epsilon_m X_m P_m g_m + P_n Y$$

where $\epsilon_m = \pm1$, $g_m \in H$ and $Y \in \mathbf{Z}[H]$. Here P_m is the unique subgroup of order p^m in H $(m = 0, \ldots, n - 1)$.

Proof. By Lemma 1.2.10,

$$A = \epsilon_0 X_0 P_0 g_0 + P_1 Y_1$$

for $\epsilon_0 = \pm 1$, $g_0 \in G$ and $Y_1 \in \mathbf{Z}[H]$. Note that $P_0 = 1$. Now suppose $n > 1$ and let $\rho : H \to H/P_1$ be the canonical epimorphism. Then $\rho(A) = \rho(P_1 Y_1) = p\rho(Y_1)$ since $\rho(X_0) = 0$ if $n > 1$ (this is not true if $X_0 = 1 + h^{2^{t-2}}$, but this case occurs only if $s = 1$ and then $n = 1$ by hypothesis).

We obtain $|\chi(\rho(Y_1)|^2 = p^{s-2}$ for every character χ on H/P_1 of order p^{t-1} since characters of order p^{t-1} on H/P_1 can be lifted to characters of order p^{t-1} on H. Again by Lemma 1.2.10, we get

$$\rho(Y_1) = \epsilon_1 X' g' + P' Y_2'$$

where $\epsilon_1 = \pm 1$, $X' = \rho(X_1)$, $g' \in H/P_1$, $Y_2' \in \mathbf{Z}[H/P_1]$, and P' is the subgroup of order p in G/P_1. Let g_1 be an element such that $\rho(g_1) = g'$. Then Lemma 1.1.13 shows

$$P_1 Y_1 = \epsilon_1 X_1 P_1 g_1 + P_2 Y_2$$

for some $Y_2 \in \mathbf{Z}[H]$, i.e.

$$A = \epsilon_0 X_0 P_0 g_0 + \epsilon_1 X_1 P_1 g_1 + P_2 Y_2.$$

The lemma follows by repeating this argument inductively. $\square$

We will use this Proposition to obtain an exponent bound for relative difference sets in Section 4.1.

A result similar to 1.2.11 is the following:

Proposition 1.2.12 (Ma, Schmidt [129]) *Let p be a prime and let $G \cong H \times P$ be an abelian group with a cyclic Sylow p-subgroup P of order p^s. Let P_i denote the unique subgroup of order p^i in G. If $Y \in \mathbf{Z}[G]$ satisfies*

$$\chi(Y) \equiv 0 \mod p^a$$

for all characters χ of order divisible by p^{s-r} where r is some fixed number $r \leq s$, $r \leq a$, then there are elements $X_0, X_1, \cdots, X_r, X$ in $\mathbf{Z}[G]$ such that

$$Y = p^a X_0 + p^{a-1} P_1 X_1 + \ldots + p^{a-r} P_r X_r + P_{r+1} X \qquad (1.12)$$

(if $r = s$ we delete the last term $P_{r+1} X$).

Proof. We write $Y = \sum_{g \in H} A_g g$ with $A_g \in \mathbf{Z}[P]$. Each element in $\mathbf{Z}[G]$ can be written like that, and, similarly, each character $\psi \in G^*$ splits into $\chi\gamma$ with $\chi \in P^*$ and $\gamma \in H^*$. We obtain

$$(\chi\gamma)(A) = \sum_{g \in H} \chi(A_g)\gamma(g)$$

or, as a matrix equation,

$$(\gamma(g))_{\gamma \in H^*, g \in H} \cdot (\chi(A_g))_{g \in H} = (a_\gamma)_{\gamma \in H^*}.$$

where $a_\gamma \equiv 0 \bmod p^a$. The matrix on the left hand side is the so called *character table* of H. The entries are units in the ring of algebraic integers. The inverse of the matrix $(\gamma(g))$ is the transpose of $(\gamma^{-1}(g))/|H|$ (this follows from the orthogonality relations (1.9)), therefore, $\chi(A_g) \equiv 0 \bmod p^a$ (note that p and $|H|$ are relatively prime) and it is enough to prove the statement for group ring elements in $\mathbf{Z}[P]$. So we may assume $Y \in \mathbf{Z}[P]$.

We define an element X_0 such that

$$\chi(X_0) = \frac{1}{p^a}\chi(Y)$$

for all characters of order p^s, hence

$$S := Y - p^a X_0$$

is an element in $\mathbf{Z}[P]$ with $\chi(S) = 0$ for all characters which are non-principal on P_1 (which are precisely the characters of order p^s). Using Corollary 1.2.5, we can write $S = P_1 T$, and we can use induction. $\square$

In the case that $r = \min\{a, s\}$, we obtain the following corollary:

Corollary 1.2.13 *If $r = \min\{a, s\}$, then*

$$Y = p^a X_0 + p^{a-1} P_1 X_1 + \ldots + p^{a-r} P_r X_r.$$

Moreover, if Y has non-negative coefficients, then we can choose the X_i such that they have non-negative coefficients.

Proof. If $r = s$ there is nothing to prove. If $r = a$ we just have to observe that P_{r+1} is a multiple of P_r.

The assertion about the non-negativity of the coefficients of the X_i is rather obvious, see also Schmidt [157]. $\square$

Proposition 1.2.12 is a generalization of the so called "Ma's Lemma". Although this lemma has been (at least implicitly) known before Ma has used it, he stated it in a very concise way:

Corollary 1.2.14 (Ma's Lemma [124]) *Let $G \cong H \times P$ be a finite abelian group with a cyclic Sylow p-subgroup of order p^t. Let χ be a character of order p^t. If $\gamma\chi(A) \equiv 0 \bmod p^r$ for all characters γ of H then we have*

$$A = p^r X + P'Y,$$

where $A, X, Y \in \mathbf{Z}[G]$ and P' is the unique subgroup of G of order p. $\square$

What can we say if $\chi(A)\overline{\chi(A)} \equiv 0$ mod p^{2i} where p is <u>not</u> the order of a cyclic Sylow p-subgroup of G? In order to be able to say something, we need the concept of "self-conjugacy".

Result 1.2.7 characterizes the Galois automorphisms which fix the prime ideal decomposition of (p) in $\mathbf{Z}[\zeta_w]$. Complex conjugation is one of these automorphisms if

$$p^j \equiv -1 \text{ mod } w'$$

(for some integer j), where $w = p^e w'$ and p does not divide w', see Lemma 1.2.15. A prime p with this property is called **self-conjugate** modulo w. In particular, p is self-conjugate modulo p^e. The concept of self-conjugacy plays an important role in the theory of difference sets in a group G. The strategy is as follows: Try to find an epimorphism ρ from G onto a "large" abelian group $\rho(G)$. Extend ρ to an epimorphism of the respective group rings or group algebras. If some self-conjugacy assumptions are satisfied, then we can perhaps conclude $\chi(\rho(D)) \equiv 0$ mod m:

Lemma 1.2.15 *Let A be an element in $\mathbf{Z}[G]$ where G is an abelian group. Let χ be a character of G of order w. If $\chi(A)\overline{\chi(A)} \equiv 0$ mod p^{2i}, then $\chi(A) \equiv 0$ mod p^i provided that p is self-conjugate modulo w.*

Proof. We use Result 1.2.7. The only case which does not follow immediately from 1.2.7 is the case where p divides w (and where the integer $w' = w/p^e$ is not divisible by p). We have $p^j + (p^e - 1)w' \equiv p^j$ mod w' and $t := p^j + (p^e - 1)w'$ is not divisible by p. If $p^j \equiv -1$ mod w', Result 1.2.7 shows that the prime ideal divisors of p in $\mathbf{Z}[\zeta_w]$ are fixed by complex conjugation.   □

We will now briefly explain the connection between character values and the eigenvalues of the corresponding group invariant matrices. Let $\mathbf{A}$ denote the group invariant matrix associated with a group ring element $A \in E[G]$, where, as before, E is a field which contains a primitive v^*-th root of unity (where v^* is again the exponent of G). Note that $\sum a_g g^{-1}$ is the group ring element corresponding to the group invariant matrix whose first row is $(a_g)_{g \in G}$. We assume that G is abelian. If χ is a character of G, we define the column vector $e_\chi := (\chi(g^{-1}))_{g \in G}$. We compute the h-entry of $\mathbf{A}e_\chi$:

$$(\mathbf{A}e_\chi)_h = \sum_{g \in G} a_{h,g}\chi(g^{-1}) = \sum_{g \in G} a_{1,gh^{-1}}\chi(hg^{-1})\chi(h^{-1})$$

which is the h-entry of e_χ multiplied by $\chi(A)$. Therefore, e_χ is an eigenvector of <u>every</u> matrix invariant under the group G. The eigenvalues are the character values of the corresponding group algebra elements. The inversion formula shows that the vectors e_χ form a basis of the group algebra $E[G]$ viewed as a vector space of dimension $|G|$ over E, actually, the inversion formula shows how to compute the group algebra element A if the eigenvalues $\chi(A)$ are known. Using the eigenvectors e_χ, we can diagonalize the algebra of group invariant matrices simultaneously! We mention all this since many of our results where character values are involved in the proof remain true if one only assumes that the incidence structure (in our situation mostly a divisible design) has an incidence matrix with an eigenvalue in an appropriate cyclotomic field. Unfortunately, I do not see how to obtain "nice" criteria when a divisible design has eigenvalues in an appropriate cyclotomic field (the only nice condition I know is that the design admits a Singer group).

The eigenvalue approach yields a nice observation: If the group G is an abelian group, then we can diagonalize the group invariant matrix $\mathbf{A}$ associated with an element $A \in K[G]$ if K contains enough roots of unity. Therefore, the K-rank is just the number of characters χ such that $\chi(A) \neq 0$. Moreover, this rank is the same as the dimension of the ideal generated by A in $K[G]$. Since we will use this later, let us state it as a proposition. Note that the K-rank of a matrix which is defined over the field K is the same as the E-rank of this matrix if K is contained in E.

Proposition 1.2.16 *Let E be a field which contains a v^*-th root of unity where v^* is the exponent of the abelian group G. The K-rank of a group invariant matrix $\mathbf{A}$ whose g, h-entry is $a_{gh^{-1}}$ equals the number of characters $\chi : K[G] \to E$ with $\chi(A) \neq 0$. (Here A is the group ring element $\sum a_g g$ corresponding to the group invariant matrix $\mathbf{A}$.) The column vector $(b_g)_{g \in G}$ is in the vector space over E generated by the columns of $\mathbf{A}$ if and only if*

$$\chi\left(\sum b_g g\right) \neq 0 \quad \Rightarrow \quad \chi\left(\sum a_g g\right) \neq 0 \tag{1.13}$$

holds for all characters.

Proof. If the vector (b_g) is in the column space of $\mathbf{A}$, then there are elements $x_h \in E$ such that

$$b_g = \sum_{h \in G} x_h a_{gh^{-1}},$$

hence

$$\sum_{g \in G} b_g g = \left(\sum_{g \in G} x_g g\right)\left(\sum_{g \in G} a_g g\right). \tag{1.14}$$

This shows that $\chi(\sum b_g g) \neq 0$ is only possible if $\chi(\sum a_g g) \neq 0$.

Now suppose that (1.13) is valid. Then we can find elements $c_\chi \in E$ such that

$$\chi\left(\sum_{g \in G} b_g g\right) = c_\chi \chi\left(\sum_{g \in G} a_g g\right).$$

Then the inversion formula shows that we can find elements $x_g \in E$ such that $\chi(\sum x_g g) = c_\chi$ for all characters. We can use these x_g's to find a linear combination of (b_g) with columns from $\mathbf{A}$ similar to (1.14). □

The vector space over a field K generated by the columns of an incidence structure is often called the **code** of the design. Thus Proposition 1.2.16 is a statement about the dimension of this code, and it gives an easy characterization of the vectors in the code. Note that we may have to extend the field K to E in order to be able to compute the characters.

Let $A = \sum a_g g$ be an element in $E[G]$, and assume that it can be written

$$A = \sum_{\chi \in G^\bullet} c_\chi e_\chi$$

with $e_\chi = \sum \chi(g^{-1})g$. The c_χ's are uniquely determined since the e_χ's form a basis of $E[G]$. We have

$$a_g = \sum_{\chi \in G^\bullet} c_\chi \chi(g^{-1}). \tag{1.15}$$

The inversion formula shows that

$$c_\chi = \chi(A)/|G|$$

is one possible choice for the c_χ, therefore the only choice since the c_χ's are unique.

This observation is of particular interest in the case of cyclic groups. Let $\mathbf{Z}_v$ be the cyclic group of order v generated by x, and let

$$A = \sum_{i=0}^{v-1} a_i x^i \tag{1.16}$$

be an element of $K[\mathbf{Z}_v]$. We can think of A as a polynomial of degree at most $v - 1$. It is straightforward to check that the group algebra $K[\mathbf{Z}_v]$ is isomorphic to the algebra $K[x]/(x^v - 1)$.

We can define a **sequence** $(b_i)_{i \in \mathbf{Z}}$ by $b_i := a_{(i \bmod v)}$. This sequence is **periodic** with period v, i.e. $b_i = b_{i+v}$. Quite often it is interesting to determine the **linear complexity** of (b_i). In several applications, as in cryptography (see Welsh [173]) or spread spectrum communication systems (see Simon, Omura, Scholtz and Levitt [160]), sequences with a large linear complexity are useful. There are several (of course equivalent) definitions of the linear complexity. For our purposes, the following definition is enough. It also explains why the complexity is sometimes called the **linear span** of a sequence: Let $(b_i)_{i \in \mathbf{Z}}$ be a sequence with period v and entries from a field K. Then the linear complexity of the sequence (b_i) is the K-dimension of the ideal generated by $\sum_{i=0}^{v-1} b_i x^i$ in $K[\mathbf{Z}_v]$. In other words, it is the rank of the circulant matrix with first column $(b_0, \ldots, b_{v-1})$. In case that the characteristic of K and v are relatively prime, we can use Proposition 1.2.16 (where G is the cyclic group of order v) to determine the linear complexity. In the cyclic case, characters are quite easy to describe: We just evaluate the polynomial (1.16) at $x = \alpha^i$, where α is a primitive v-th root of unity in E. Choosing different exponents i yields different characters.

Sometimes, the sequence (b_i) is <u>defined</u> via

$$b_i = \sum_{j=0}^{v-1} c_j (\alpha^i)^j \tag{1.17}$$

where α is a primitive v-th root of unity in E and the c_j's are suitable elements of E. We will see some examples of sequences defined in this way later in this monograph. Although the c_j's and α are elements in E, it is possible that the entries b_i in the sequence are always members of a subfield K of E which does <u>not</u> contain α. What is the linear complexity of the sequence (b_i)? We have just observed that the characters of the cyclic group $\mathbf{Z}_v$ generated by x are the mappings

$$\chi_j : x^i \longmapsto (\alpha^{-j})^i.$$

We obtain

$$b_i = \sum_{j=0}^{v-1} c_j \chi_j(x^{-i}).$$

If we compare this expression with (1.15), we get

$$c_j = \frac{1}{v} \cdot \chi_j \Big(\sum_{i=0}^{v-1} b_i x^i \Big).$$

In the (important) case that E is a finite field, we obtain the following result:

Corollary 1.2.17 (Key [107]) *Let q be a prime power and v a positive integer which divides $q - 1$, i.e., the field $GF(q)$ contains a primitive v-th root of unity α. Let $(b_i)_{i \in \mathbf{Z}}$ be a sequence with entries in a subfield K of E defined via (1.17). Then the linear complexity of the sequence (b_i) is $|\{j : 0 \le j \le v - 1, c_j \ne 0\}|$.*

1.3 Multipliers

In the preceeding section, we have seen that it is sometimes interesting to obtain information about $\chi(D)$ if $\chi(D)\overline{\chi(D)}$ is known. We can say something if prime ideal divisors of $\chi(D)$ are fixed by complex conjugation. However, this is a purely algebraic phenomenon. We do not need more information about D or the equation which D satisfies (see (1.7)). Using this equation, it is sometimes possible to say more: We can find more automorphisms which fix the ideals $(\chi(D))$. If we know, for instance, that $D^{(t)} = Dg$ holds for some t relatively prime to $|G|$, then $(\chi(D))$ is obviously fixed by the Galois automorphism $\zeta_w \to \zeta_w^t$, where ζ_w is a primitive complex w-th root of unity, and χ is a character of order w. An integer t relatively prime to $|G|$ with the property $D^{(t)} = Dg$ for some $g \in G$ is called a **(numerical) multiplier** of D. More generally, multipliers are group automorphisms φ such that $\varphi(D) := \{\varphi(a) : a \in D\}$ is a translate Dg of D. Loosely speaking, multipliers are those group automorphisms which induce automorphisms of the design corresponding to the difference set D. The most important multipliers are the numerical multipliers since they yield directly Galois automorphisms fixing $\chi(D)$. Integers t (which are not necessarily multipliers) with this property are called χ-multipliers of A: More precisely, an integer t is a χ-**multiplier** of a difference set D in an abelian group if the ideal generated by $\chi(D)$ is fixed by the Galois automorphism $\zeta_w \to \zeta_w^t$. Here χ

denotes a complex character of order w and ζ_w a primitive w-th root of unity. Note that t is necessarily relatively prime to w. If t is a χ-multiplier it is not necessarily true that t is also a multiplier of D, see the remarks preceeding Proposition 2.4.18.

Let T be an automorphism group of the design corresponding to a difference set in G. The multipliers in T together with the Singer group G generate the normalizer of G in T:

Theorem 1.3.1 (Bruck [38]) *Let G be a Singer group of the design $\mathcal{D}$ and let M denote the group of multipliers in T where T is an automorphism group of $\mathcal{D}$ which contains G. Then the semidirect product MG is the normalizer of G in T.*

Proof. To distinguish between the points of $\mathcal{D}$ (which are group elements in G) and the group G acting on $\mathcal{D}$, we denote the design automorphism $x \mapsto xg$ by τ_g. Let δ be a multiplier. Then $\delta^{-1}\tau_g\delta = \tau_{\delta(g)}$, hence M is in the normalizer of G. Now let δ be in the normalizer N of G in T such that $\delta(0) = 0$ (otherwise replace δ by $\delta\tau_h$ for a suitable $h \in G$). The mapping $\delta^{-1}\tau_g\delta$ maps 0 to $\delta(g)$, hence $\delta^{-1}\tau_g\delta = \tau_{\delta(g)}$. But this mapping is an automorphism of G, hence δ is a multiplier. $\qquad\square$

We call two difference sets A and B **multiplier equivalent** if there is a multiplier φ such that $\varphi(A) = Bg$ for some translate of B. Multipliers are very useful in the theory of difference sets since we know several multiplier theorems. A "multiplier theorem" shows that certain group automorphisms have to be multipliers. If it is known that an integer (or a group automorphism) has to be a multiplier, one can use this multiplier to construct the difference set or to prove that it cannot exist. Before we illustrate some applications of multipliers, we state and sketch a proof of the so called "second multiplier theorem". The proof illustrates how useful algebraic number theory can be to prove theorems about difference sets.

Theorem 1.3.2 (2nd multiplier theorem) *Let D be a (v, k, λ)-difference set in an abelian group G of exponent v^*. Let t be an integer relatively prime to v and let n_1 be a divisor of $n := k - \lambda$. Let $n_1 = p_1{}^{e_1} \cdots p_s{}^{e_s}$ be the prime factorization of n_1 and $n_2 := n_1/(v, n_1)$. For each p_i, we define*

$$q_i = \begin{cases} p_i & \text{if } p_i \text{ does not divide } v \\ l_i & \text{if } v^* = p_i{}^r u_i, \ (p_i, u_i) = 1. \text{ Here } l_i \text{ is an integer} \\ & \quad \text{such that } (l_i, p_i) = 1 \text{ and } l_i \equiv p_i{}^f \ mod \ u_i. \end{cases}$$

For each i, we assume the existence of an integer f_i and a multiplier s_i such that $s_iq_i{}^{f_i} \equiv t \ mod \ v^*$. If $n_2 > \lambda$ or if the equation $FF^{(-1)} = (n/n_2)^2$ in $\mathbf{Z}[G]$ has only the trivial solution $F = (n/n_2)g$, then t is a multiplier of D.

Proof. We have $\chi(D)\chi(D^{(-1)}) = n$ for all non-principal characters since D is a difference set which satisfies $DD^{(-1)} = n + \lambda G$. The Galois automorphism

$$\zeta \to \zeta^x \tag{1.18}$$

fixes the ideal generated by $\chi(D)$ if $x = s_i$ since s_i is a multiplier (here ζ is a primitive complex v^*'th root of unity, and all our calculations are carried out in $\mathbf{Z}[\zeta]$). On the other hand, the ideal generated by p_i for a prime divisor of n_1 is fixed by the automorphism in (1.18) if $x = q_i{}^{f_i}$. (We have to quote Result 1.2.7 (b) if p_i does not divide v and part (c) if $v^* = p_i{}^r u_i$.) Therefore, we know that $\zeta \to \zeta^t$ fixes the ideal which "divides" both $(\chi(D))$ and (n_1). This shows

$$\chi(D^{(t)} D^{(-1)}) \equiv 0 \bmod n_1$$

for all non-principal characters and implies

$$D^{(t)} D^{(-1)} = n_2 F + \lambda G \tag{1.19}$$

with $\chi_0(F) = n/n_2$ for a suitable element $F \in \mathbf{Z}[G]$, see Corollary 1.2.5. The coefficients of $DD^{(t)}$ are non-negative, hence the coefficients of F are non-negative if $n_2 > \lambda$. Now we compute

$$
\begin{aligned}
(D^{(t)} D^{(-1)})(DD^{(-t)}) &= (n_2 F + \lambda G)(n_2 F^{(-1)} + \lambda G) \\
&= (n + \lambda G)(n + \lambda G)
\end{aligned}
$$

which shows

$$FF^{(-1)} = \left(\frac{n}{n_2}\right)^2 .$$

If this equation has only the trivial solution $F = (n/n_2)g$ for a suitable $g \in G$, then t is a numerical multiplier (we just multiply (1.19) by D to get $nD^{(t)} = nDg$). But this is the only solution at least if the coefficients of F are non-negative, hence if $n_2 > \lambda$. □

Corollary 1.3.3 (first multiplier theorem, Hall, Ryser [81]) *Let D be an abelian $(v, k, \lambda; n)$-difference set of order n, and let p be a prime divisor of n which is relatively prime to v. If $p > \lambda$ then p is a multiplier of D.* □

It is widely conjectured that any divisor of the order of an abelian difference set which is relatively prime to the group order has to be a multiplier. However, in all <u>known</u> proofs of multiplier theorems, a technical condition like "$n_2 > \lambda$" is needed.

Problem 3 (multiplier conjecture) Is every divisor of the order of an abelian difference set in G (relatively prime to $|G|$) a multiplier?

In connection with the multiplier conjecture, the following corollary of Theorem 1.3.2 is of interest:

Corollary 1.3.4 *Let D be an abelian $(v, k, \lambda; n)$-difference set where $n = p^a$ is a power of the prime p and $(p, v) = 1$. Then p is a multiplier of D.*

Proof. We may assume $k \leq v/2$ (otherwise we replace D by its complement). Then our basic equation (1.3) implies

$$\lambda = k(k-1)/(v-1) \leq k((v/2) - 1)/(v-1) < k/2$$

which shows $n > \lambda$. Then Theorem 1.3.2 shows that p is a multiplier using $n_1 = n$.   □

Multipliers have been introduced by Hall [79]. He proved a multiplier theorem for cyclic planar difference sets. Later, this theorem has been generalized to arbitrary symmetric designs (Hall and Ryser [81], Bruck [38], and Hall [80]). The second multiplier theorem is basically due to Menon [136]. The version above is in Arasu and Xiang [22]. In most multiplier theorems, it is assumed that the integer n_1 is relatively prime to n. The slight generalization above uses an observation which is contained in Ganley and Spence [73] and in Lam [114]. Sufficient conditions when the equation $FF^{(-1)} = (n/n_2)^2$ has only the "trivial" solution have been obtained by McFarland [133] and Lander [116]. Recently, these have been slightly improved by Qiu [152]. For other proofs of the multiplier theorem we refer to Lander [116] and Pott [141], [143].

What can we say about multipliers of other (symmetric) incidence structures? Multiplier theorems for divisible difference sets are in Arasu and Xiang [22] and Ko and Ray-Chaudhuri [110]. Later in this monograph, we will prove a multiplier theorem for an incidence structure which is a slight generalization of a divisible design (see Theorem 5.3.5). Here we will state a multiplier theorem for relative difference sets whose proof is rather analogous to the proof of Theorem 1.3.2 (and is therefore omitted). It is basically contained in [22]. Preliminary versions have been obtained by Elliott and Butson [68] and Lam [114].

Theorem 1.3.5 *Let R be an (m, n, k, λ)-difference set in an abelian group G of exponent v^*. Let t be an integer relatively prime to v and let k_1 be a divisor of k. Let $k_1 = p_1{}^{e_1} \cdots p_s{}^{e_s}$ be the prime factorization of k_1 and $k_2 := k_1/(mn, k_1)$. For each p_i, we define*

$$q_i = \begin{cases} p_i & \text{if } p_i \text{ does not divide } v \\ l_i & \text{if } v^* = p_i{}^r u_i, \ (p_i, u_i) = 1. \text{ Here } l_i \text{ is an integer} \\ & \text{such that } (l_i, p_i) = 1 \text{ and } l_i \equiv p_i{}^f \mod u_i. \end{cases}$$

For each i, we assume the existence of an integer f_i and a multiplier s_i such that $s_i q_i{}^{f_i} \equiv t \mod v^*$ If $k_2 > \lambda$ or if the equation $FF^{(-1)} = (k/k_2)^2$ in $\mathbf{Z}[G]$ has only the trivial solution $F = (k/k_2)g$, then t is a multiplier of R, provided that t is a multiplier of the underlying $(m, k, n\lambda)$-difference set.   □

Corollary 1.3.6 *Let R be an abelian relative difference set with parameters*

$$\left(\frac{q^{d+1} - 1}{q - 1}, n, q^d, \frac{q^d - q^{d-1}}{n} \right)$$

where $q = p^a$ is a prime power and $(n, p) = 1$. Then p is a multiplier of R.

Proof. Note that p is a multiplier of the underlying difference set in view of Corollary 1.3.4. We put $k_2 = q^d$ which proves the corollary. $\qquad\square$

Multipliers of (relative) difference sets R are useful since it is known that there are translates Rg which are fixed by multipliers. The following result is basically contained in Elliott and Butson [68]. The proof is quite analogous to the corresponding result on difference sets. We will see later that this result is not true in the semiregular case. But first, we need a well-known lemma:

Result 1.3.7 (orbit theorem [29, Chapter III]) *Let $\mathcal{D}$ be a regular square design. Then an automorphism φ of $\mathcal{D}$ has the same number of fixed points as fixed blocks. If G is an automorphism group of $\mathcal{D}$, then the number of point orbits of G equals the number of orbits on blocks.*

Theorem 1.3.8 *Let R be a regular relative (m, n, k, λ)-difference set with multiplier τ. Then there is at least one translate such that $\tau(Rg) = Rg$. If mn and k are relatively prime, and if the difference set is abelian, then there is at least one translate fixed by __all__ multipliers.*

Proof. The orbit theorem on automorphisms of regular designs shows that there is a translate fixed by a multiplier since each multiplier fixes at least one point (the identity element in the group). In order to prove the second statement, we consider the difference set R and a translate Rg. Let α be the product of the elements in R, then the product of the elements in Rg is αg^k. Since k is relatively prime to the group order, there is a unique translate of R such that the product of its elements is 1. This translate is fixed by all multipliers. $\qquad\square$

Example 1.3.9 We are now going to illustrate some applications of the multiplier theorem. We try to find a difference set with parameters $(11, 5, 2)$. In this case, $n = 3$ is a multiplier. We may assume that our putative difference set D is fixed by this multiplier. If $x \in D$, then $3x$, $9x$, $5x$ and $4x$ are also elements in D. We choose $x = 1$ and see that $\{1, 3, 4, 5, 9\}$ is indeed a difference set in the cyclic group of order 11. Now let us try to construct a $(43, 7, 1)$-difference set. Since 43 is a prime, we may assume that 1 is an element of D. Since 2 is a multiplier, we get that $2, 4, 8, 16, 32, 21, 42, \ldots$ have to be elements in D, which is impossible since $|D| = 7$.

We will finish this section with a new multiplier theorem on relative difference sets:

Theorem 1.3.10 (Pott, Reuschling, Schmidt [151]) *Let R be a relative $(m, d, m - 1, (m - 2)/d)$-difference set in an abelian group G. Then $m - 1$ is a multiplier of R.*

Proof. We replace R by a translate Rg, $g \in G$, if necessary, such that $R \cap N = \emptyset$. We set

$$t = \left(\prod_{h \in N} h \right)^{\frac{m-2}{d}} .$$

Note that $\prod_{h \in N} h$ is an involution or 1, hence $t^2 = 1$.

For $g \in G \setminus N$, let $\gamma(g)$ be the unique element of R in Ng. Let r be an arbitrary element of R and define

$$t' := \prod_{\substack{r_1, r_2 \in R \\ r_1 r_2^{-1} \in Nr}} r_1 r_2^{-1} .$$

We are now going to calculate t' in two different ways. From the definition of a relative difference set it is immediate that

$$t' = tr^{m-2} .$$

On the other hand, we have

$$t' = \prod_{\substack{r' \in R \\ r' \neq r}} r' \gamma(r^{-1} r')^{-1} = r^{-1} \gamma(r^{-1})$$

as $\gamma(r^{-1} r')$ ranges over $R \setminus \{\gamma(r^{-1})\}$ if r' ranges over $R \setminus \{r\}$. Hence

$$r^{m-1} = r^{m-2} r = t' t^{-1} r = \gamma(r^{-1}) t$$

for all $r \in R$. Thus $R^{(m-1)} = Rt$.   □

We note that this theorem is <u>not</u> a consequence of Theorem 1.3.5:

Example 1.3.11 Theorem 1.3.5 shows that 9, 17 and 25 are multipiers of every abelian relative $(16, 2, 15, 7)$-difference set: We take $k_1 = 15$, $p_1 = 3$, $p_2 = 5$. Then $9 \equiv 3^2 \equiv 5^6 \bmod 32$ and $25 \equiv 5^2 \equiv 3^6 \bmod 32$ as required. But it is not possible to get the multipiers 3 and 5 using Theorem 1.3.5 since no power of 3 is congruent 5 mod 32 and no power of 5 is congruent 3 mod 32. But Theorem 1.3.8 shows that 3 and 5 actually are multipliers of every abelian relative $(16,2,15,7)$-difference set using the multiplier 15 in Theorem 1.3.5: We have $3 \equiv 3 \bmod 32$ and $3 \equiv 15 \cdot 5^7 \bmod 32$.

This example can be generalized. Note that the underlying difference set of a relative $(m, d, m-1, (m-2)/d)$-difference set is a trivial $(m, m-1, m-2)$-difference set, hence every integer is a numerical multiplier of the underlying difference set.

Corollary 1.3.12 *Let R be an abelian relative $(m, d, m-1, (m-2)/d)$-difference set. If $m - 1 = p^i q^j$ is the product of two prime powers, then p^i and q^j are both multipliers of R.*

Proof. Let k be the order of q modulo mn (note that p and q are relatively prime to mn). Then we use Theorem 1.3.5 with $k_1 = m - 1$. We have

$$
\begin{aligned}
p^i &\equiv p^i \bmod mn, \\
p^i &\equiv (m - 1)q^{k-j} \bmod mn
\end{aligned}
$$

which proves the corollary. □

Chapter 2

Examples: Existence and non-existence

In the last chapter, we have summarized the algebraic tools which are needed in order to investigate group ring or group algebra equations $AA^{(-1)} = M$. Here we restrict ourselves to the case that A is a group ring element with 0 and 1 coefficients which describes a divisible design. Of course, one can apply the theory to any other incidence structure (as long as the number of points equals the number of blocks). One can take, for instance, generalized quadrangles with a Singer group. The "classical" difference sets are, of course, also divisible difference sets and we will summarize the known series of difference sets in Section 2.1. Although this monograph is not mainly intended to be a monograph about difference sets, it is worthwhile to summarize at least the known families of difference sets (in most cases without proof), in particular, since relative difference sets are extensions of difference sets in the usual sense. In Section 2.2, we describe the known series of relative difference sets, and Section 2.3 contains several interesting series of divisible difference sets. In the final Section 2.4, we will develop non-existence results on divisible difference sets. These non-existence results provide some well-known non-existence results about usual difference sets as well.

2.1 Difference sets

In this section, we state most of the known series of difference sets. Our list is not quite complete. First, it is possible that difference sets with the same **parameters** as those listed here exist but in different groups. Second, there are a few more examples of the so called cyclotomic difference sets (see Result 2.1.9) which are not mentioned here. We refer the reader to the books of Baumert [26] and Storer [165] for a complete list of known cyclotomic difference sets.

Recall that an (m, k, λ)-difference set is a k-subset of a group G of order m such that the list of non-zero differences (or quotients in the multiplicative notation) contains each non-zero group element exactly λ times. The order of the difference set is $n := k - \lambda$ and hence we will also speak about $(m, k, \lambda; n)$-difference sets. The order of a difference set is important since a difference set D satisfies the group ring equation

$$DD^{(-1)} = n + \lambda G$$

in $\mathbf{Z}[G]$, hence if we apply a non-principal character χ to this equation we get

$$\chi(D)\overline{\chi(D)} = n$$

and we can use algebraic number theory.

We begin with the most famous class of difference sets which is due to Singer [161]. The *trace*-function is the usual trace function from $\mathrm{GF}(q^{d+1})$ onto $\mathrm{GF}(q)$ which maps β to $\sum_{i=0}^{d} \beta^{q^i}$. The trace function depends on the size of the extension field and the "ground field". If it is not clear from the context, which trace we are looking at we write $\mathrm{trace}_{q^{d+1}/q}$ (or $\mathrm{trace}_{e/f}$ if we consider the trace from $\mathrm{GF}(q^e)$ onto $\mathrm{GF}(q^f)$), otherwise we omit the subscripts. Sometimes in this monograph (as in the construction of Singer difference sets below) we will use results about finite fields. The reader is referred to the books by Jungnickel [101] and Lidl and Niederreiter [120].

Theorem 2.1.1 *Let q be a prime power and d a non-negative integer. Let α be a generator of the multiplicative group of $\mathrm{GF}(q^{d+1})$ (such an element is called primitive). Then the set of integers $\{i : 0 \le i < (q^{d+1}-1)/(q-1), \mathrm{trace}(\alpha^i) = 0\}$ modulo $(q^{d+1} - 1)/(q-1)$ forms a (cyclic) difference set with parameters*

$$\left(\frac{q^{d+1} - 1}{q - 1}, \frac{q^d - 1}{q - 1}, \frac{q^{d-1} - 1}{q - 1}; q^{d-1} \right). \tag{2.1}$$

*The designs corresponding to these so called **Singer difference sets** are the point-hyperplane designs of the projective geometry $PG(d, q)$.*

Proof. The point-hyperplane design of $PG(d, q)$ consists of the 1-dimensional subspaces of $\mathrm{GF}(q)^{d+1}$ as points and the subspaces of codimension 1 as blocks. We identify the vector space $\mathrm{GF}(q)^{d+1}$ with the field $\mathrm{GF}(q^{d+1})$, hence the multiplicative group $\mathrm{GF}(q^{d+1})^*$ is identified with the non-zero elements of $\mathrm{GF}(q)^{d+1}$. Then the points of $PG(d, q)$ are the elements of the quotient group

$$\mathrm{GF}(q^{d+1})^*/\mathrm{GF}(q)^*.$$

The mapping

$$T_\alpha : \quad \mathrm{GF}(q^{d+1})^* \quad \rightarrow \quad \mathrm{GF}(q^{d+1})^*$$
$$v \quad \mapsto \quad v\alpha$$

induces a bijection of $\mathrm{GF}(q^{d+1})^*/\mathrm{GF}(q)^*$ onto itself. Let us call this induced map T_α again. Now T_α has order $(q^{d+1} - 1)/(q - 1)$ and it is easy to see that T_α is an automorphism of $PG(d, q)$. Therefore, the symmetric design $PG(d, q)$

admits a cyclic automorphism group which acts regularly on points and blocks (note that transitivity on points implies transitivity on blocks, see Result 1.3.7). This shows the existence of a cyclic difference set with the desired parameters. In order to find the difference set, we choose α as our base point, and the base block is the set of points with trace$(\beta) = 0$. To be more precise: We have to choose the coset of GF$(q)^*$ containing α as the base point, and the base block consists of those cosets whose elements have 0-trace (if one element in a coset has 0-trace, then the trace of all elements in the coset is 0). Since GF$(q)^* = \langle \alpha^{(q^d-1)/(q-1)} \rangle$, we can restrict ourselves to ith powers of α with $i < (q^d - 1)/(q - 1)$ in order to describe GF$(q^{d+1})^*/$GF$(q)^*$. □

Example 2.1.2 The smallest example of a Singer difference set is $\{1, 2, 4\}$ in $\mathbb{Z}_7$ corresponding to PG$(2, 2)$, the projective plane of order 2. Another example, which does not correspond to a projective plane, is the $(15, 7, 3)$-difference set $\{0, 1, 2, 4, 5, 8, 10\}$ corresponding to PG$(4, 2)$. In order to construct this difference set, we have to find a primitive element (or a primitive polynomial) in GF(16). (We call a polynomial of degree n over GF(q) *primitive* if a root of it is primitive, hence if it has order $q^n - 1$.) We take α to be a root of $x^4 + x^3 + 1$. The following table shows the values of trace$_{16/2}(\alpha^i)$:

i	0	1	2	3	4	5	6	7	8	9	10	11	12	13	14
trace(α^i)	0	0	0	1	0	0	1	1	0	1	0	1	1	1	1

It shows that $\{0, 1, 2, 4, 5, 8, 10\}$ is the desired difference set. If we choose another primitive element, we obtain another (but equivalent) difference set.

We refer the reader to Section 3.1 for more about the classical Singer difference sets. Now let us continue our description of difference sets. The next series is due to McFarland [134].

Theorem 2.1.3 *Let q be a prime power and d a positive integer. Let G be an abelian group of order $v = q^{d+1}(q^d + \ldots + q^2 + q + 2)$ which contains an elementary abelian subgroup E of order q^{d+1}. View E as the additive group of GF$(q)^{d+1}$. Put $r = (q^{d+1} - 1)/(q - 1)$ and let $H_1, \ldots, H_r$ be the hyperplanes (i.e., the linear subspaces of dimension d) of E. If $g_1, \ldots, g_r$ are distinct coset representatives of E in G, then $D = (g_1 + H_1) \cup (g_2 + H_2) \cup \ldots \cup (g_r + H_r)$ is a difference set with parameters*

$$\left(q^{d+1}(1 + \frac{q^{d+1} - 1}{q - 1}), \frac{q^d(q^{d+1} - 1)}{q - 1}, \frac{q^d(q^d - 1)}{q - 1}; q^{2d} \right). \tag{2.2}$$

□

Several variations of McFarland's construction are contained in Dillon [63], which yield non-abelian examples as well. We may ask whether there are other groups (where the Sylow p-subgroup is not elementary abelian) which contain McFarland difference sets. In case that $q = p^f$ and p is self conjugate modulo the order of G, this is impossible:

Result 2.1.4 (Ma, Schmidt [128]) *Let G be a group containing a McFarland difference set with parameters (2.2). If p is an odd prime which is self-conjugate modulo the order of G, and if the Sylow p-subgroup P of G is an abelian normal subgroup of G, then P has to be elementary abelian.*

If $p = 2$, the following difference set with the McFarland parameters is known which does not follow from Theorem 2.1.3. Up to now, this example is sporadic, i.e. it is not a member of an infinite series. It shows that Result 2.1.4 cannot hold for $p = 2$.

Example 2.1.5 (Arasu, Sehgal [21]) The elements

(0000), (0001), (0002), (0020), (0101), (0031), (0132), (0200), (0212), (0220),
(0221), (0302), (1012), (1101), (1132), (1201), (1211), (1222), (1231), (1322)

in $\mathbf{Z}_2 \times \mathbf{Z}_4 \times \mathbf{Z}_4 \times \mathbf{Z}_3$ form a $(96, 20, 4)$-difference set.

The following variation of the McFarland construction is due to Spence [164]:

Theorem 2.1.6 *Let E be the elementary abelian group of order 3^{d+1} and let G be a group of order $v = 3^{d+1}(3^{d+1} - 1)/2$ containing E. Put $m = (3^{d+1} - 1)/2$ and let $H_1, \ldots, H_m$ denote the subgroups of E of order 3^d. If $g_1, \ldots, g_m$ are distinct coset representatives of E in G, then $D = (g_1 + (E \backslash H_1)) \cup (g_2 + H_2) \cup (g_3 + H_3) \cup \ldots \cup (g_m + H_m)$ is a* **Spence difference set**. *It has parameters*

$$\left(\frac{3^{d+1}(3^{d+1} - 1)}{2}, \frac{3^d(3^{d+1} + 1)}{2}, \frac{3^d(3^d + 1)}{2}; 3^{2d} \right).$$

□

In Section 2.3 we will show that the McFarland and Spence construction indeed yield difference sets: It turns out that a variation of their constructions gives rise to divisible difference sets.

If $q = 2$, the parameters of McFarland difference sets are

$$(2^{2d+2}, 2^{2d+1} - 2^d, 2^{2d} - 2^d; 2^{2d}).$$

These are a special type of difference sets called **Hadamard difference sets**: A difference set in G whose order is $|G|/4$ is called a Hadamard difference set (other authors use the name **Menon difference set**. More generally, we call a symmetric $(m, k, \lambda; n)$-design an **H-design** if $n = m/4$. (We do not call these designs "Hadamard designs" since this term is used by most authors for designs with $m = 4n - 1$, which are also closely related to Hadamard matrices.) It is well known that the order of a Hadamard difference set (or an H-design) has to be a square (see Corollary 1.1.6) since $|G|$ is even, hence we have $m = 4u^2$. Using the basic equation (1.3) we get $k = 2u^2 \pm u$ and we can write the parameters of a Hadamard difference set as

$$(4u^2, 2u^2 \pm u, u^2 \pm u; u^2).$$

There are many constructions of these difference sets. The following abelian examples are known:

Result 2.1.7 *Let G be an abelian group of order $4u^2$ with $u = 2^a 3^b w^2$ where w is the product of not necessarily distinct primes $\equiv 3 \bmod 4$. If*

$$G \cong \mathbf{Z}_{2^{a_1}} \times \ldots \times \mathbf{Z}_{2^{a_s}} \times \mathbf{Z}_{3^{b_1}}^2 \times \ldots \times \mathbf{Z}_{3^{b_r}}^2 \times EA(w^2)$$

with $\sum a_i = 2a + 2$ ($a \geq 0$, $a_i \leq a + 2$), $\sum b_i = 2b$ ($b \geq 0$), then G contains a Hadamard difference set of order u^2. (Note: $EA(w^2)$ denotes the group of order w^2 which is the direct product of groups of prime order.)

It is beyond the scope of this monograph to give a proof of this result but we refer the reader to the survey of Davis and Jedwab [53]. Result 2.1.7 includes quite a few sufficient conditions on the existence of Hadamard difference sets. In Section 2.4 we will discuss the quality of these conditions (necessary = sufficient?).

It is quite interesting that there are many non-abelian Hadamard difference sets which are completely "different" from the abelian ones. To be more precise: Several constructions of non-abelian difference sets are known which use an abelian difference set in G to start with and try to find a point regular group H within the group generated by G and the multiplier group. Of course, we get a difference set in H which is isomorphic to the original abelian difference set. Smith [162] calls a non-abelian difference set **genuinely non-abelian** if the corresponding design has no abelian point-regular group acting on it. Some examples are known, see Kibler [108] and Smith [162]. In case of Hadamard difference sets, examples of non-abelian difference sets are known where no abelian difference set with the same parameters can exist (of course, these difference sets are genuinely non-abelian). The first example of a non-abelian $(100, 45, 20)$-difference set has been constructed by Smith [163]: Result 2.4.14 shows that no abelian difference set with these parameters can exist.

Example 2.1.8 (Smith [163]) The following element in $\mathbf{Z}[G]$ describes a $(100, 45, 20)$-difference set in $G := \langle a, b, c : a^5 = b^5 = c^4 = 1, ab = ba, aca^{-1} = a^2 c, bcb^{-1} = b^2 c \rangle$:

$$(1 + a + a^4) + (1 + a)b + (1 + a^2 + a^3 + a^4)b^2 + (1 + a + a^2 + a^3)b^3$$
$$+(1 + a^4)b^4 + [(a^2 + a^4) + a^4 b + a^3 b^2 + (1 + a^2)b^3 + (a + a^2 + a^3 + a^4)b^4]c$$
$$+[a^4 + (a + a^2 + a^4)b + (a + a^4)b^2 + (1 + a^2 + a^4)b^3 + a^3 b^4]c^2$$
$$+[(a^3 + a^4) + (1 + a^4)b + a^3 b^2 + (a + a^2 + a^3 + a^4)b^3 + ab^4]c^3.$$

Known recursive constructions (see Corollary 2.2.4) yield many more non-abelian Hadamard difference sets starting with this example.

We will finish this little survey on difference sets with the so called cyclotomic difference sets. Many difference sets have been constructed as subgroups (or unions of cosets of subgroups) of the multiplicative group of a finite field viewed as a subset of the additive group. The most famous examples are the Paley difference sets consisting of the squares modulo q for a prime power $q \equiv 3 \bmod 4$. But there are more constructions. We refer the reader to Baumert [26] and Storer

[165] for a proof of the following result (and a few more examples, see also Section 6.3).

Result 2.1.9 *Let $q = ef + 1$ be a prime power. We define*

$$C_i^{(e)} := \{x^j : x \in GF(q)\backslash\{0\}, j \equiv i \ mod \ e\}.$$

Then the following sets are difference sets in the additive group of $GF(q)$:

(a) $C_0^{(2)}$, $q \equiv 3 \ mod \ 4$ *(quadratic residues,* **Paley difference sets***);*

(b) $C_0^{(4)}$, $q = 4t^2 + 1, t$ *odd* **biquadratic residues***;*

(c) $C_0^{(4)} \cup \{0\}$, $q = 4t^2 + 9, t$ *odd;*

(d) $C_0^{(8)}$, $q = 8t^2 + 1 = 64u^2 + 9, t, u$ *odd;*

(e) $C_0^{(8)} \cup \{0\}$, $q = 8t^2 + 49 = 64u^2 + 441, t$ *odd, u even;*

(f) $C_0^{(6)} \cup C_1^{(6)} \cup C_3^{(6)}$, $q = 4t^2 + 27, (t, 3) = 1$ (**Hall difference sets**).

Let q and $q + 2$ be prime powers. Then the set $D = \{(x, y) : x, y$ are both non-zero squares or both non-squares or $y = 0\}$ is a **twin prime power difference set** *with parameters*

$$\left(q^2 + 2q, \frac{q^2 + 2q - 1}{2}, \frac{q^2 + 2q - 3}{4}; \frac{q^2 + 2q + 1}{4}\right).$$

All these difference sets are called **cyclotomic difference sets***.*

2.2 Relative difference sets

We have seen already that relative difference sets are extensions of "ordinary" difference sets. We will begin this section with a list of parametrically all known examples. Relative (m, n, k, λ)-difference sets exist at least in the following cases:

1. $(m, k, n\lambda) = (m, m, m)$, $m = 4u^2$ or $8u^2$, $n = 2$ and there exists a Hadamard difference set of order u^2;

2. $(m, k, n\lambda) = (m, m, m)$ and m is a prime power;

3. $(m, k, n\lambda) = (\frac{q^{d+1}-1}{q-1}, q^d, q^{d-1})$, $n|(q - 1)$;

4. $(m, k, n\lambda) = (\frac{q^{d+1}-1}{q-1}, q^d, q^{d-1})$, $n = 2$, d odd;

5. $(m, k, n\lambda) = (21, 16, 12)$, $n|6$.

Now let us look at these cases separately. The first two series are semiregular relative difference sets, and we begin our investigation with this case. It turns out that a <u>splitting</u> relative $(m, 2, m, m/2)$-difference set exists in $G \times N$ relative to a subgroup N of order 2 if and only if a Hadamard difference set in G exists:

Proposition 2.2.1 (Jungnickel [95]) *Let R be a splitting relative $(m, 2, m, m/2)$-difference set in $G \times N$ relative to N. Then $R \cap G$ is a Hadamard difference set in G. Conversely, if D is a Hadamard difference set in G ($|G| = 4u^2$), then*

$$\{(d, 1) : d \in D\} \cup \{(e, t) : e \notin D\} \subset G \times \{1, t\}$$

is a relative difference set with parameters $(4u^2, 2, 4u^2, 2u^2)$ (where $\{1, t\}$ is the forbidden subgroup N).

Proof. If D is a Hadamard difference set, it is easy to check that the construction yields a relative difference set. Conversely, we put $D = R \cap G$ and compute

$$DD^{(-1)} + (G - D)(G - D)^{(-1)} = \frac{|G|}{2} + \frac{|G|}{2}G$$

(these are the difference representations in G). We get

$$2DD^{(-1)} - 2|D|\, G + |G|\, G = \frac{|G|}{2} + \frac{|G|}{2}G$$

which shows that D is a difference set of order $|G|/4$, and thus D is a Hadamard difference set. □

Corollary 2.2.2 *Splitting relative difference sets with parameters*

$$(4u^2, 2, 4u^2, 2u^2) \tag{2.3}$$

exist whenever u^2 is the order of a Hadamard difference set (see Result 2.1.7).□

It is also possible to construct non-splitting relative $(m, 2, m, m/2)$-difference sets. For this purpose we need the following recursive construction. It is basically contained in Davis [49], the slightly more general statement below is in Pott [148] (but only for the abelian case).

Lemma 2.2.3 *Let G be a group of order $m_1 m_2 n$ containing a subgroup N of order n. Let G_1 and G_2 be subgroups of G of order $m_1 n$ and $m_2 n$ with*

$$G_1 \cap G_2 = N \tag{2.4}$$

If G_1 contains an $(m_1, n, m_1, m_1/n)$-difference set R_1 relative to N and G_2 an $(m_2, n, m_2, m_2/n)$-difference set R_2 relative to N, then

$$R_1 R_2 = \{r_1 r_2 : r_1 \in R_1, r_2 \in R_2\}$$

is an $(m_1 m_2, n, m_1 m_2, m_1 m_2/n)$-difference set relative to N provided that G_1 and G_2 commute, i.e. $gh = hg$ for all elements $g \in G_1$ and $h \in G_2$.

Proof. We use group ring notation and compute $R_1 R_2 R_1^{(-1)} R_2^{(-1)}$ in $\mathbf{Z}[G]$. First of all we note that $R_1 R_2$ has $0/1$-coefficients because of (2.4). We obtain

$$
\begin{aligned}
R_1 R_2 R_1^{(-1)} R_2^{(-1)} &= \left(m_1 + \frac{m_1}{n}(G_1 - N) \right) \cdot \left(m_2 + \frac{m_2}{n}(G_2 - N) \right) \\
&= m_1 m_2 + \frac{m_1 m_2}{n}(G_2 - N) + \frac{m_1 m_2}{n}(G_1 - N) + \\
&\quad + \frac{m_1 m_2}{n^2}(G_1 - N)(G_2 - N) \\
&= m_1 m_2 + \frac{m_1 m_2}{n}(G_1 + G_2 - 2N) + \\
&\quad + \frac{m_1 m_2}{n^2}(nG - nG_1 - nG_2 + nN) \\
&= m_1 m_2 + \frac{m_1 m_2}{n}(G - N).
\end{aligned}
$$

$\square$

Corollary 2.2.4 *If G and H are two groups which contain Hadamard difference sets then $G \times H$ contains a Hadamard difference set.* $\square$

We can use the construction in Lemma 2.2.3 with a (splitting) relative $(4u^2, 2, 4u^2, 2u^2)$-difference set and the (non-splitting) cyclic relative $(2, 2, 2, 1)$-difference set $\{0, 1\}$ in $\mathbf{Z}_4$ to get <u>non-splitting</u> relative difference sets:

Corollary 2.2.5 (Arasu, Jungnickel, Pott [12]) *Relative difference sets with parameters*

$$(8u^2, 2, 8u^2, 4u^2) \tag{2.5}$$

relative to N exist in $G \times \mathbf{Z}_4$ whenever G contains a Hadamard difference set and where N is contained in a complement of G. $\square$

We leave it to the reader to verify that the construction in Corollary 2.2.5 yields difference sets no translates of which are fixed by the multiplier

$$
\begin{aligned}
\varphi: \quad G \times \mathbf{Z}_4 &\rightarrow \quad G \times \mathbf{Z}_4 \\
(g, h) &\mapsto \quad (g, -h).
\end{aligned}
$$

This shows that the orbit theorem (Result 1.3.7) does not hold for semiregular designs.

It is impossible that splitting relative difference sets with parameters (2.5) exist since Proposition 2.2.1 shows that in that case a group of order $8u^2$ would contain a Hadamard difference set, which is impossible. However, there is a **Hadamard matrix** of size $8u^2$ but which is not group invariant. Let us be more specific. A Hadamard matrix of order v is a $(v \times v)$-matrix with entries ± 1 which satisfies $\mathbf{HH}^t = v\mathbf{I}_v$. It is well-known that the order of a Hadamard matrix is 1, 2 or it is divisible by 4, see Beth, Jungnickel and Lenz [29], for instance. The three smallest examples are

$$(1), \quad \begin{pmatrix} 1 & 1 \\ 1 & -1 \end{pmatrix}, \quad \begin{pmatrix} 1 & -1 & -1 & -1 \\ -1 & 1 & -1 & -1 \\ -1 & -1 & 1 & -1 \\ -1 & -1 & -1 & 1 \end{pmatrix}.$$

It is conjectured that Hadamard matrices of order v exist whenever v is divisible by 4. We refer the reader to the recent survey by Seberry and Yamada [159].

People have also studied **generalized Hadamard matrices**. These are matrices $\mathbf{H} = (h_{i,j})$ of size m with entries from a group N of order n such that

$$\mathbf{HH}^* = m\mathbf{I} + \frac{m}{n}\mathbf{GJ} :$$

The (i,j)-entry in $\mathbf{H}^*$ is defined $h_{j,i}^{-1}$ and we view $\mathbf{H}$ and $\mathbf{H}^*$ as matrices with entries in $\mathbf{Z}[N]$. The equation above means that the list of quotients $h_{i,j}h_{i',j}$, $j = 1, \ldots, m$, contains every element of N exactly m/n times $(i \neq i')$. If $N = \{-1, 1\}$ is a cyclic group of order 2, and $\mathbf{H}$ is considered as a matrix over the integers, then $\mathbf{H}$ is just a usual Hadamard matrix. Now let us replace each entry $h_{i,j}$ in the generalized Hadamard matrix by the corresponding group invariant $n \times n$-matrix to get a matrix $\mathbf{A}$. This matrix is the incidence matrix of a divisible $(m, n, m, 0, m/n)$-design. If $\mathbf{H}$ is group invariant (with group G), then the matrix $\mathbf{A}$ is group invariant with group $G \times N$ and N stabilizes the point classes of this design. In particular, there is a splitting relative $(m, n, m, m/n)$-difference set in $G \times N$ relative to N.

If we have $m = n$ then the matrix $\mathbf{A}$ describes an $(n, n, n, 0, 1)$-design which can exist if and only if a projective plane of order n exists, see Section 5.4. The most challenging open problem about generalized Hadamard matrices is the following:

Problem 4 Let $\mathbf{H}$ be a generalized Hadamard matrix of size m with entries from a group of order m. Is it true that m has to be a prime power?

It has been an open problem for quite a while whether there are generalized Hadamard matrices over groups N which are not elementary abelian. First examples where N is not elementary abelian have been constructed by de Launey [57], see also Result 4.2.3.

It is not yet known whether m has to be a prime power even if we assume that $\mathbf{H}$ is group invariant! We will obtain some partial results in Section 5.4. Moreover, we refer the reader to Dey and Hayden [62] for a recent investigation of generalized Hadamard matrices.

If we multiply rows or columns of a Hadamard matrix by -1, the result is a Hadamard matrix, again. Therefore, we can *normalize* a Hadamard matrix such that the first row and column consist of +1's. For instance, the matrix

$$\mathbf{H} = \begin{pmatrix} 1 & -1 & -1 & -1 & 1 & -1 & -1 & -1 \\ -1 & 1 & -1 & -1 & -1 & 1 & -1 & -1 \\ -1 & -1 & 1 & -1 & -1 & -1 & 1 & -1 \\ -1 & -1 & -1 & 1 & -1 & -1 & -1 & 1 \\ 1 & -1 & -1 & -1 & -1 & 1 & 1 & 1 \\ -1 & 1 & -1 & -1 & 1 & -1 & 1 & 1 \\ -1 & -1 & 1 & -1 & 1 & 1 & -1 & 1 \\ -1 & -1 & -1 & 1 & 1 & 1 & 1 & -1 \end{pmatrix}$$

is a Hadamard matrix of order 8; the normalized version is

$$\mathbf{H'} = \begin{pmatrix} 1 & 1 & 1 & 1 & 1 & 1 & 1 & 1 \\ 1 & 1 & -1 & -1 & 1 & 1 & -1 & -1 \\ 1 & -1 & 1 & -1 & 1 & -1 & 1 & -1 \\ 1 & -1 & -1 & 1 & 1 & -1 & -1 & 1 \\ 1 & 1 & 1 & 1 & -1 & -1 & -1 & -1 \\ 1 & 1 & -1 & -1 & -1 & -1 & 1 & 1 \\ 1 & -1 & 1 & -1 & -1 & 1 & -1 & 1 \\ 1 & -1 & -1 & 1 & -1 & 1 & 1 & -1 \end{pmatrix}.$$

If we delete the first row and column from $\mathbf{H'}$ and replace -1 by 0, we obtain the incidence matrix of a symmetric $(7, 3, 1)$-design. This is, by no means, an incidence:

Proposition 2.2.6 *Let $\mathbf{H}$ be a normalized Hadamard matrix of order $4n$. If we delete the first row and column from $\mathbf{H}$ and replace -1 by 0, we obtain an incidence matrix $\mathbf{M}$ of a symmetric design with parameters*

$$(4n - 1, 2n - 1, n - 1; n). \tag{2.6}$$

Proof. Since $\mathbf{H}$ is normalized, there are precisely $2n - 1$ entries 1 in each row of $\mathbf{M}$. The inner product of two distinct rows e_1 and e_2 of $\mathbf{M}$ is the number of positions where e_1 as well as e_2 have an entry 1. In order to determine this number, we look at the inner product of the corresponding rows e_1' and e_2' in $\mathbf{H}$. We define four numbers $a_{i,j}$, $i, j = \pm 1$: $a_{i,j}$ denotes the number of positions where e_1' has the entry i and e_2' has entry j. We obtain

$$\begin{aligned} a_{1,1} + a_{1,-1} &= 2n, \\ a_{1,1} + a_{-1,1} &= 2n, \\ a_{1,1} + a_{-1,-1} &= 2n, \\ a_{1,-1} + a_{-1,1} &= 2n. \end{aligned}$$

The first two equations hold since $\mathbf{H}$ is normalized. It follows that $a_{1,1} = n$, which shows that the "λ"-value of the symmetric design is $n - 1$. $\qquad\square$

In some cases, the designs constructed in Proposition 2.2.6 have a difference set representation. This is the reason why many people call difference sets with parameters (2.6) Hadamard difference sets. Since we have used this term already, we will call difference sets with parameters (2.6) **Paley-Hadamard difference sets** (which is not the standard terminology). The reason is that the Paley difference sets (squares in GF(q)) are the most prominent members in this series. Other Paley-Hadamard difference sets are the twin primes and the Singer difference sets corresponding to PG($d, 2$).

What is the connection between Hadamard difference sets and Hadamard matrices? A Hadamard matrix is called **regular** if it has constant row sum. Not every Hadamard matrix can be transformed into a regular matrix by multiplying some rows and columns with -1. If we replace -1 by 0 in a regular Hadamard matrix **H** of order v, we get the incidence matrix of a symmetric (v, k, λ)-design. To get a (non-trivial) equation which combines v, k and λ, we take two distinct rows of **H**. The inner product is 0, on the other hand it is $\lambda + (v - 2k + \lambda) - (k - \lambda - (k - \lambda))$. This shows

$$v = 4(k - \lambda)$$

and hence the parameters are those of an H-design, in particular $v = 4u^2$. (Our Hadamard matrix of size 8 above cannot be regular since 8 is not of the form $4u^2$.) If this design has a point regular group, i.e., if the regular Hadamard matrix is group invariant, the matrix yields a Hadamard difference set. Conversely, every H-design gives rise to a regular Hadamard matrix.

We have seen that there are two types of designs associated with Hadamard matrices: We can construct a symmetric $(4n-1, 2n-1, n-1; n)$-design from every Hadamard matrix. If the matrix is regular, we find symmetric $(4u^2, 2u^2 - u, u^2 - u; u^2)$-designs. No parameters are known for which an H-design exists but no Hadamard difference set (although not all H-designs come from difference sets). For a thorough investigation of Hadamard matrices and their designs, we refer to Assmus and Key [24]. What is the connection between Hadamard matrices and relative $(m, 2, m, m/2)$-difference sets? In the splitting case, Proposition 2.2.1 gives an answer. In the non-splitting case, we know the following:

Proposition 2.2.7 *Let $\mathcal{D}$ be a divisible semiregular $(m, n, m, m/n)$-design which admits an automorphism group N of order n acting regularly on each point class. (In particular, this is true if $\mathcal{D}$ can be desribed by a normal difference set.) Then there exists a generalized Hadamard matrix of size m with entries in N.*

Proof. From each of the m point classes, we choose a base point p_i. Moreover, let $B_1, \ldots, B_m$ denote the blocks through some point, say p_1. The matrix whose i, j-entry $h_{i,j}$ is the group element in N which maps the point p_i to a point on B_j is a generalized Hadamard matrix. Firts of all note that $h_{i,j}$ is uniquely determined. Two points p_i and $h(p_{i'})$ $(i \neq i')$ are joined by exactly m/n blocks. A block $g(B_j)$ $(g \in N)$ joins p_i and $h(p_{i'})$ if and only if $h_{i,j} = h_{i',j}h$, hence there are exactly m/n column indices j with $h_{i',j}^{-1}h_{i,j} = h$. If we replace the entries in H by their invereses, we get a generalized Hadamard matrix. $\square$

In the splitting case, we can label the rows and columns of the generalized Hadamard matrix constructed above in such a way that the matrix is group invariant.

Corollary 2.2.8 *The existence of a Hadamard matrix of order m is equivalent to the existence of an $(m, 2, m, m/2)$-design.*   □

This corollary shows, in particular, that m is always divisible by 4. There are many non-existence results about abelian relative $(m, 2, m, m/2)$-difference sets known, however, as far as I know, there is no value of m for which it is known that <u>no</u> relative difference sets with the parameters above can exist.

Problem 5 Find new constructions or non-existence results for relative $(m, 2, m, m/2)$-difference sets.

Up to now, we have considered only very "small" liftings of trivial (m, m, m)-difference sets to relative difference sets, namely those where the order of the forbidden subgroup is 2. But there are more extensions. Parametrically, in all these examples m is a prime power p^a. They can be realized (parametrically) as projections from relative $(p^a, p^a, p^a, 1)$-difference sets which we are now going to construct. Relative $(p^a, p^a, p^a, 1)$-difference sets correspond to projective planes with a certain type of quasiregular automorphism group, see Chapter 5. Using this description, the following two theorems are rather obvious, see also Dembowski and Ostrom [60].

If p is odd, there is a very easy construction:

Theorem 2.2.9 *Let G be the additive group of the field $GF(q)$ where q is odd. Then the set*

$$R := \{(x, x^2) : x \in GF(q)\} \subset G \times G \tag{2.7}$$

is a $(q, q, q, 1)$-difference set relative to $\{0\} \times G$.

Proof. No element in $G \times \{0\}$ has a difference representation with elements from R. If another element has two representation $(x - y, x^2 - y^2) = (x' - y', x'^2 - y'^2)$ with $x \neq y$ and $x' \neq y'$, then $x + y = x' + y'$. This shows $2x = 2x'$, hence $x = x'$ since q is odd. Because of the cardinality of R and G, R is a relative difference set with the desired parameters.   □

The case $q = 2^a$ is slightly more complicated. In order to construct a relative $(q, q, q, 1)$-difference set with even q, we need some preparations. We define the <u>set</u>

$$H = \{(x, y) : x, y \in \mathrm{GF}(q)\}.$$

This set becomes a group isomorphic to $\mathbf{Z}_4^a$ with the multiplication $(x, y) \oplus (w, z) := (x + w, y + z + xw)$. We have $\ominus(x, y) = (x, y + x^2)$.

Theorem 2.2.10 *Suppose $q = 2^a$. Then the subset*

$$R := \{(x, x) : x \in GF(q)\} \tag{2.8}$$

of $(H, \oplus)$ is a $(2^a, 2^a, 2^a, 1)$-difference set relative to $N := \{(0, x) : x \in GF(q)\}$.

Proof. It is again obvious that no element in N has a difference representation with elements from R. If $(x, x) \ominus (y, y) = (x', x') \ominus (y', y') \neq (0, 0)$, we get

$$(x + y, x + y^2 + y + xy) = (x' + y', x' + y'^2 + y' + x'y').$$

This is only possible if $x = x'$ and $y = y'$. □

Corollary 2.2.11 *Abelian relative difference sets with parameters*

$$(p^a, p^b, p^a, p^{a-b})$$

exist for all primes p and integers a and b with $a \geq b$. □

We will investigate these difference sets in more detail in Section 5.4. Parametrically, Corollary 2.2.11 contains a list of all relative difference sets with $n \neq 2$ (this case has to be considered separately) which are extensions of (m, m, m)-difference sets. But there are many more constructions in other groups, and these will be investigated in Chapter 4.

The next series of relative difference sets is an analogue of the Singer difference sets.

Theorem 2.2.12 (Bose [31]; Elliott, Butson [68]) *Let q be a prime power and let α be a primitive element of $GF(q^d)$. Then there exist cyclic relative difference sets with parameters*

$$\left(\frac{q^d - 1}{q - 1}, q - 1, q^{d-1}, q^{d-2}\right) \tag{2.9}$$

in the multiplicative group of $GF(q^d)$. They consist of the powers of α which satisfy $\text{trace}(\alpha^i) = 1$. The designs corresponding to these relative difference sets are the divisible designs consisting of the non-zero points in $GF(q)^d$ with the affine hyperplanes not through the origin as blocks.

Proof. It is obvious that the multiplicative group of the field generated by α acts sharply transitively on the points of the divisible design described above. The set of elements whose trace is 1, is one of the blocks of that design. □

We call difference sets with parameters (2.9) **affine difference sets**. If they are constructed via the trace function as described above, they are called the **classical affine difference sets**. Similar to the Singer difference set case, there are non-classical examples, see Chapter 3.2.

The projections of the affine difference sets are difference sets with the parameters of the complements of the Singer difference sets. (The reader should be careful: The parameters d in the affine and the projective case differ by 1.) Obviously, the classical affine difference sets project onto the complements of the classical Singer difference sets. If q is odd, the order of the forbidden subgroup is even. It is therefore always possible to find a relative difference set whose forbidden subgroup has order 2 (via projection) and which is an extension of the complement of a Singer difference set. If q is even, there is another extension which will be described in Section 3.3.

Theorem 2.2.13 *Cyclic relative difference sets with parameters*

$$\left(\frac{q^d - 1}{q - 1}, 2, q^{d-1}, \frac{q^{d-1} - q^{d-2}}{2} \right) \tag{2.10}$$

exist whenever q is a power of 2 and d is odd.

Proof. See Theorem 3.3.4.

It is worth mentioning that the only <u>known</u> non-trivial difference sets which admit liftings to relative difference sets have the parameters of the complements of the Singer difference sets. Most of them have parameters

$$\left(\frac{q^d - 1}{q - 1}, n, q^{d-1}, \frac{q^{d-1} - q^{d-2}}{n} \right) \tag{2.11}$$

where n divides $q - 1$. One example where n is not a divisor of $q - 1$ is the Series in Theorem 2.2.13. Only one more example is known.

Example 2.2.14 (Lam [114]) The set

$$\{0, 2, 4, 5, 816, 17, 32, 54, 64, 73, 83, 90, 97, 103, 108\} \subset \mathbf{Z}_{126}$$

is a cylic relative $(21, 6, 16, 2)$ difference set. The underlying difference set is the complement of the classical Singer difference set with parameters $(21, 16, 12)$.

The following two questions arise quite naturally:

Problem 6 Is it true that the only non-trivial difference sets which have extensions to relative difference sets have parameters

$$\left(\frac{q^d - 1}{q - 1}, q^{d-1}, q^{d-1} - q^{d-2} \right) \tag{2.12}$$

with a prime power q?

Problem 7 Do relative (m, n, k, λ)-difference sets exist such that the projection is a difference set with parameters (2.12) but where $n \neq 2$ and where n is not a divisor of $q - 1$ (except the $(21, 6, 16, 2)$-difference set in Example 2.2.14)?

The reader who wants to construct a relative difference set with certain (known) parameters can find a construction in this section. However, there are many more inequivalent constructions and many examples in other groups.

2.3 Divisible difference sets

Many series of divisible difference sets are known, and I suppose that there will be many more series which are not yet known: The definition of a divisible difference set seems to be much weaker than the definition of a relative difference set. It is not our aim to summarize all known series in this section. We restrict ourselves to three cases which are of particular interest:

- We construct divisible $(m, n, k, \lambda_1, \lambda_2)$-difference sets with $mn = 4(k - \lambda_2)$. These examples will be of interest in view of Theorem 6.2.1.

- We construct semiregular divisible difference sets. The existence of these difference sets seems to be rare. It is, in my opinion, interesting to investigate the semiregular divisible difference sets since one of the most interesting classes of <u>relative</u> difference sets are semiregular difference sets.

- We will consider divisible difference sets which generalize the McFarland and Spence series of difference sets (which proves Theorems 2.1.3 and 2.1.6).

We begin with divisible $(m, n, k, \lambda_1, \lambda_2)$-difference sets with property

$$mn = 4(k - \lambda_2). \tag{2.13}$$

Property (2.13) generalizes the defining property for Hadamard difference sets. The following recursive construction generalizes a well-known recursive construction for Hadamard difference sets, see Corollary 2.2.4.

Theorem 2.3.1 (Arasu, Pott [15]) *Let D be a Hadamard difference set with parameters $(4u^2, 2u^2 - u, u^2 - u)$ in G, and let T be a divisible $(m, n, k, \lambda_1, \lambda_2)$-difference set in H relative to N. Then the set*

$$E := \{(x, y) : x \in D \text{ and } y \in T, \text{ or } x \notin D \text{ and } y \notin T\} \subset G \times H$$

is a divisible difference set relative to $\{1\} \times N$ if and only if T has property (2.13) or $m = 1$. In this case, the set E has property (2.13), and it has parameters $(m', n, k', \lambda_1', \lambda_2')$ with

$$
\begin{aligned}
m' &= 4u^2 m, \\
k' &= 2u^2 mn + umn - 2uk, \\
\lambda_1' &= 4u^2 \lambda_1 + 2u^2 mn - 4u^2 k - 2uk + umn, \\
\lambda_2' &= u^2 mn - 2uk + umn.
\end{aligned}
$$

Proof. The difference sets D and T satisfy the equations

$$DD^{(-1)} = u^2 + (u^2 - u)G$$

and

$$TT^{(-1)} = (k - \lambda_1) + (\lambda_1 - \lambda_2)N + \lambda_2 H.$$

We write
$$E = [D, T] + [G - D, H - T]$$
(this notation is self-explanatory) and calculate $EE^{(-1)}$:

$$
\begin{aligned}
EE^{(-1)} &= [DD^{(-1)}, TT^{(-1)}] + 2[D(G-D)^{(-1)}, T(H-T)^{(-1)}] \\
&\quad + [(G-D)(G-D)^{(-1)}, (H-T)(H-T)^{(-1)}] \\
&= [u^2 + (u^2 - u)G, (k - \lambda_1) + (\lambda_1 - \lambda_2)N + \lambda_2 H] \\
&\quad + 2[-u^2 + u^2 G, (k - \lambda_2)H - (k - \lambda_1) - (\lambda_1 - \lambda_2)N] \\
&\quad + [u^2 + (u^2 + u)G, (k - \lambda_1) + (\lambda_1 - \lambda_2)N + (mn - 2k + \lambda_2)H].
\end{aligned}
$$

Let x_i $(i = 1, 2, 3, 4)$ be the coefficients of $(G - 1, H - N)$, $(G - 1, N - 1)$, $(1, H - N)$, and $(G - 1, 1)$, respectively, in $EE^{(-1)}$. We obtain

$$
\begin{aligned}
x_1 &= (u^2 + u)mn - 2uk, \\
x_2 &= (u^2 + u)mn - 2uk, \\
x_3 &= (u^2 + u)mn - 2uk + u^2(mn - 4(k - \lambda_2)), \\
x_4 &= (u^2 + u)mn - 2uk.
\end{aligned}
$$

This shows that E is a divisible difference set in $G \times H$ relative to $\{1\} \times N$ if and only if T has property (2.13) or if $H = N$, i.e., if $m = 1$: In the latter case, the exceptional value x_3 vanishes since the "set" $[1, H - N]$ is empty. $\square$

Corollary 2.3.2 (Jungnickel [95]) *Let D be a Hadamard difference set of order u^2 in G and let T be an (n, k, λ)-difference set in N. Then there exists a divisible difference set with parameters*

$$(4u^2, n, 2u^2 n - 2ku + un, 2u^2 n + un - 2uk - 4u^2 k + 4u^2 \lambda, u^2 n + un - 2ku)$$

in $G \times N$ relative to N. $\square$

Example 2.3.3 We can construct, for instance, divisible difference sets with parameters
$$(4, 3, 5, 1, 2), \quad (4, 5, 7, 3, 2) \quad \text{and} \quad (4, 6, 8, 4, 2).$$
choosing the trivial $(n, 1, 0)$-difference set.

It is a quite natural question to ask which relative difference sets have property (2.13). Here is the answer:

Proposition 2.3.4 (Arasu, Pott [15]) *A relative (m, n, k, λ)-difference set R satisfies $mn = 4(k - \lambda)$ with $n \geq 2$ if and only if $n = 2$ and R is an extension of a trivial difference set.*

Proof. From $mn = 4(k - \lambda)$ we conclude $k^2 - \lambda mn = (k - 2\lambda)^2$. The basic equation for relative difference sets is $k^2 - k = \lambda(mn - n)$, thus $k^2 - \lambda mn = k - \lambda n$. We obtain (using $n \geq 2$)

$$(k - 2\lambda)^2 = k - \lambda n \leq k - 2\lambda \leq (k - 2\lambda)^2.$$

We conclude that $n = 2$ and either $k = n\lambda$ or $k = n\lambda + 1$, which shows that R projects onto a trivial difference set. □

Other divisible difference sets with $mn = 4(k - \lambda_2)$ and $n = 2$ are known:

Corollary 2.3.5 *Divisible difference sets with parameters*

$$(4u^2(q + 1), 2, 4u^2(q + 1) + 2u, 4u^2 + 2u, 2u^2(q + 1) + 2u)$$

and property (2.13) exist whenever q is an odd prime power and u^2 is the order of a Hadamard difference set.

Proof. Use the construction in Theorem 2.3.1 together with $(q + 1, 2, q, 0, (q - 1)/2)$-difference sets (projections of the classical affine difference sets). □

If we choose a $(2^a, 2, 2^a, 0, 2^{a-1})$-difference set in Theorem 2.3.1, then we get a non-splitting difference set with parameters $(2^{a+2}u^2, 2, 2^{a+2}u^2, 0, 2^{a+1}u^2)$, which we can also obtain using Lemma 2.2.3.

We will now construct a few semiregular divisible difference sets. It is easy to see that our construction in Theorem 2.3.1 yields semiregular divisible difference sets if T is semiregular. So we begin our brief summary of semiregular divisible difference sets with those having property (2.13):

Theorem 2.3.6 *Let q be a prime power and s an integer with $1 \leq s \leq n$. Divisible difference sets with parameters*

$$\left(\frac{q^{2n-2s}(q^s - 1)}{q - 1}, q^s, \frac{q^{2n-s-1}(q^s - 1)}{q - 1}, \frac{q^{2n-s-1}(q^{s-1} - 1)}{q - 1}, \frac{q^{2n-s-2}(q^s - 1)}{q - 1} \right)$$

exist in any abelian group G which contains $EA(q^n)$ as a subgroup. These difference sets are always semiregular.

Proof. Let V be the vector space $\mathrm{GF}(q)^n$, and let L be an arbitrary s-dimensional subspace. We define $r := q^{n-s}(q^s - 1)/(q - 1)$. Let G be a group of order rq^n containing V (to be precise $EA(q^n)$) as a subgroup. Let $g_1, \ldots, g_r$ be coset representatives of V in G with $g_1 \in V$. Let $H_1, \ldots, H_r$ be the hyperplanes, i.e., $(n - 1)$-dimensional subspaces of V which do not contain L. There are exactly r such hyperplanes (note that $r = (q^n - 1)/(q - 1) - (q^{n-s} - 1)/(q - 1)$ is the difference between the number of all hyperplanes and the number of hyperplanes containing L). Our divisible difference set will be the union of the $g_i H_i$ or, in group ring notation, $R = \sum g_i H_i$. Let us calculate $RR^{(-1)}$ in $\mathbf{Z}[G]$. We have to use the basic facts

$$
\begin{aligned}
H_i^2 &= q^{n-1} H_i, & (2.14) \\
H_i H_j &= q^{n-2} V \quad (i \neq j). & (2.15)
\end{aligned}
$$

It follows

$$
\left(\sum g_i H_i\right)\left(\sum g_i^{-1} H_i\right) = q^{n-1} \sum H_i + r q^{n-2} \sum_{i \neq 1} g_i V
$$

since every element $\rho(g_i)$, $i \neq 1$, has exactly r representations $\rho(g_k)\rho(g_j)^{-1}$, $k \neq j$, where ρ denotes the canonical epimorphism from G onto G/V. What can we say about $\sum H_i$? We have

$$
\sum_{i=1}^{r} H_i = r + \gamma_1 (L - 0) + \gamma_2 (V - L)
$$

with

$$
\begin{aligned}
\gamma_1 &= \frac{q^{n-s}(q^{s-1} - 1)}{q - 1}, \\
\gamma_2 &= \frac{q^{n-s-1}(q^s - 1)}{q - 1}.
\end{aligned}
$$

Note that γ_1 is just the number of hyperplanes in an $(n - 1)$-dimensional vector space not containing a fixed $(s - 1)$-dimensional subspace. Similarly, γ_2 is the number of hyperplanes in an $(n - 1)$-dimensional space that does not contain a given s-dimensional subspace. Therefore $\gamma_2 q^{n-1} = r q^{n-2}$ which shows that the coefficients of the elements not in L in the product RR^{-1} is a constant $\lambda_2 = r q^{n-2}$. The coefficients of the non-zero elements in L is $\gamma_1 q^{n-1} = \lambda_1$. □

The case $s = 1$ is contained in Davis [50]. On the opposite side of the range, if $s = n$, we obtain the series (5.11) in Jungnickel [95]. If $s = n$ and $q = 2$, the difference sets satisfy (2.13). Recently, Davis and Jedwab found another example of semiregular divisible difference sets which we are now going to describe.

Theorem 2.3.7 (Davis, Jedwab [54]) *Let q be a prime power. We assume the existence of an abelian $(m, (q^n - 1)/(q - 1), \lambda)$-difference set $T = \{t_i : i = 1, \ldots, (q^n - 1)(q - 1)\}$ in M and an elementary abelian (q, k, μ)-difference set D in $EA(q)$. Then there exists a divisible difference set with parameters*

$$
\left(m, q^n, \frac{kq^{n-1}(q^n - 1)}{q - 1}, \frac{kq^{n-1}(q^n - 1)}{q - 1} - q^{2(n-1)}(k - \mu), q^{n-2}k^2\lambda \right)
$$

in $E := M \times EA(q^n)$ relative to $EA(q^n)$.

Proof. Let H_i be a hyperplane in the vectorspace $V = GF(q)^n$ whose additive group is denoted G. The additive group of V/H_i contains a (q, k, μ)-difference set D. The canonical epimorphism $V \to V/H_i$ is denoted by φ_i. We define D_i to be the preimage $\varphi_i^{-1}[D]$ of D in G. These subsets (or the corresponding group ring elements) satisfy

$$
\begin{aligned}
D_i D_i^{(-1)} &= q^{n-1}\mu G + (k - \mu)q^{n-1}H_i, \\
D_i D_j^{(-1)} &= q^{n-2}k^2 G,
\end{aligned}
$$

see (2.14) and (2.15). The group ring element

$$
R := \sum_{i=1}^{(q^n-1)(q-1)} t_i D_i \quad \text{in } \mathbf{Z}[E]
$$

"is" the desired divisible difference set:

$$
\begin{aligned}
RR^{(-1)} &= \sum_i t_i t_i^{-1} D_i D_i^{-1} + \sum_{i \neq j} t_i t_j^{-1} D_i D_j^{-1} \\
&= \frac{q^{n-1}\mu(q^n - 1)}{q - 1}G + (k - \mu)q^{n-1}\sum_i H_i + \lambda q^{n-2}k^2(E - G).
\end{aligned}
$$

This shows already $\lambda_2 = \lambda q^{n-2}k^2$. It is straightforward to compute the λ_1-value using

$$
\sum_{i=1}^{(q^n-1)/(q-1)} H_i = \frac{q^n - 1}{q - 1} + \frac{q^{n-1} - 1}{q - 1}G. \tag{2.16}
$$

□

Corollary 2.3.8 *If T is a trivial $((q^d-1)/(q-1), (q^d-1)/(q-1), (q^d-1)/(q-1))$-difference set then the difference sets constructed in Theorem 2.3.7 are semiregular.*

□

We note that there are many examples of difference sets in elementary abelian groups (cyclotomic difference sets) and there are examples of difference sets whose k-value is $(q^n - 1)/(q - 1)$ (Singer difference sets). These difference sets can be used in Theorem 2.3.7.

In [51], Davis introduced the term **almost difference sets**: An almost difference set is a divisible difference set with parameters $(m, n, k, \lambda_1, \lambda_2)$ and $|\lambda_1 - \lambda_2| = 1$, i.e., they are "almost" difference sets. We have seen already examples of almost difference sets: Relative difference sets with $\lambda = 1$ provide examples, a difference set with $\lambda_1 \neq 0$ is in Example 1.1.10. Davis obtained more examples which are also semiregular. We state the parameters without proof. The construction is actually a slight modification of a construction of some Hadamard difference sets in Arasu, Davis, Jedwab and Sehgal [6].

Result 2.3.9 (Davis [51]) *Let H be a group of order 4 and $G \cong H \times \mathbf{Z}_{3^a}^2$. Then G contains an almost difference set with parameters*

$$(4, 3^{2a}, 2 \cdot (3^{2a} - 3^a), 3^{2a} - 2 \cdot 3^a, 3^{2a} - 2 \cdot 3^a + 1)$$

relative to $\mathbf{Z}_{3^a}^2$. These difference sets are semiregular.

Another construction of Davis generalizes the McFarland construction of difference sets, it gives examples of almost difference sets and of semiregular divisble difference sets. We are now going to decribe this construction.

Theorem 2.3.10 (Davis [51]) *Let $D = \{d_1, \ldots, d_h\}$ be an $(m, n, h, h - 1, \lambda_2)$-divisible difference set in M relative to N where $h = (q^{d+1} - 1)/(q - 1)$. Let G denote the additive group of the vector space $V = GF(q)^{d+1}$. If H_i $(i = 1, \ldots h)$ are the hyperplanes of V, then the set R corresponding to the group ring element*

$$R = \sum_{i=1}^{n} d_i H_i$$

is a divisible difference set with parameters

$$\left(m, nq^{d+1}, \frac{q^d(q^{d+1} - 1)}{q - 1}, \frac{q^d(q^d - 1)}{q - 1}, q^{d-1}\lambda_2 \right)$$

in $M \times G$ relative to $N \times G$. Moreover, R is semiregular if and only if $d = 1$ and D is semiregular.

Proof. We compute

$$
\begin{aligned}
RR^{(-1)} &= \sum_{i=1}^{h} d_i H_i \sum_{j=1}^{h} d_j^{(-1)} H_j \\
&= q^d \left(\frac{q^{d+1} - 1}{q - 1} + \frac{q^d - 1}{q - 1} G \right) \\
&\quad + q^{d-1}[(h - 1)G(N - 1) + \lambda_2 G(M - N)]
\end{aligned}
$$

and see that R becomes a divisible difference set if

$$\frac{q^d(q^d - 1)}{q - 1} = q^{d-1}(h - 1).$$

If this is true, then the trouble term G vanishes. This is the reason why we had to require "$k - \lambda_1 = 1$". It is straightforward to verify the parameters of the divisible difference set.

The difference set R is semiregular if and only if $h^2 = mn\lambda_2$ which is satisfied if and only if D is semiregular. $\square$

We refer the reader to Arasu, Jungnickel and Pott [14] for a thorough investigation of divisible difference sets with $k - \lambda_1 = 1$. Trivial examples are the sets $(M \setminus N) \cup \{1\}$ with parameters

$$(m, n, (m-1)n + 1, (m-1)n, (m-1)n - n + 2). \tag{2.17}$$

and trivial difference sets with $n = 1$:

$$(m, 1, m - 1, m - 2, m - 2). \tag{2.18}$$

If we choose the trivial examples (2.18), we get exactly the McFarland series. We can get almost difference sets at most if $d = 1$. If this is the case and if we take difference sets with parameters (2.17) and $n = 1$ or $n = 3$, we get almost difference sets (if $n = 3$, then q has to be a power of 3).

Another variation of the McFarland construction is the following Theorem. Here we call a difference set (in the usual sense) **antisymmetric** if it satisfies $D + D^{(-1)} + 1 = G$ (see also Result 5.2.14).

Theorem 2.3.11 (Arasu, Pott [18]) *Let q be a prime power and d a positive integer. Let V be the elementary abelian group $EA(q^d)$ of order q^d and let G be any group of order $v = 1 + (q^d - 1)/(q - 1)$. Let D be a (v, k, λ)-difference set in G with $1 \notin D$. Assume that either $k = 0$ or $k = v - 1$ or D is an antisymmetric difference set. Then there exists a divisible difference set in $G \times V$ relative to V with parameters*

$$\left(1 + \frac{q^d - 1}{q - 1}, q^d, q^{d-1}(v - r - 1) + k(q^d - q^{d-1}), \lambda_1, \lambda_2\right)$$

with

$$\lambda_1 = kq^d - 2kq^{d-1} + \frac{q^{2d-2} - q^{d-1}}{q - 1},$$

$$\lambda_2 = \frac{q^{2d-2} - q^{d-1}}{q - 1} + \delta(q^{d-1} - 2q^{d-2}) + \lambda(q^d - 4q^{d-1} + 4q^{d-2})$$

where $\delta = 0$ if $k = 0$, $\delta = 2k - 2$ if $k = v - 1$ and $\delta = 2k - 1$ otherwise.

Proof. Define $h = (q^d - 1)/(q - 1)$ and let H_i $(i = 1, \ldots, h)$ denote (as ususal) the hyperplanes in V where we think of V as the additive group of $\mathrm{GF}(q^d)$. Write $D = \{g_1, \ldots, g_k\}$ and let g_0 be the identity element in G. Then

$$R = \sum_{i=1}^{k} g_i(V - H_i) + \sum_{i=k+1}^{v-1} g_i H_i$$

describes the desired divisible difference set. We leave the details to the reader (it should be rather routine). □

If $k = 0$, then $\lambda_1 = \lambda_2$, and R is a McFarland difference set, see Theorem 2.1.3. A slight variation yields the Spence difference sets:

Theorem 2.3.12 (Arasu, Pott [18]) *Let q be a prime power and d a positive integer. Let $V = EA(q^d)$ and let G be any group of order $h = (q^d-1)/(q-1)$. Let $D = \{d_1, \ldots, d_k\}$ be a (v, k, λ)-difference set in G Then there exists a divisible difference set in $G \times V$ with parameters*

$$\left(\frac{q^d - 1}{q - 1}, q^d, q^{d-1}(v - k) + k(q^d - q^{d-1}), \lambda_1, \lambda_2 \right)$$

with

$$\lambda_1 = \frac{q^{2d-2} - q^{d-1}}{q - 1} + kq^d - 2kq^{d-1},$$

$$\lambda_2 = \frac{q^{2d-2} - q^{d-1}}{q - 1} + 2k(q^{d-1} - 2q^{d-2}) + \lambda(q^d - 4q^{d-1} + 4q^{d-2}).$$

Proof. Using the notation as in Theorem 2.3.11, we define

$$R = \sum_{i=1}^{k} g_i(V - H_i) + \sum_{i=k+1}^{v} g_i H_i.$$

This is the divisible difference set.   □

In this case, we obtain difference sets if and only if $v = k+1$ and $q = 3$. This is the Spence construction of difference sets, see Theorem 2.1.6.

The reader has seen some recursive constructions and also many constructions using hyperplanes in $\mathrm{GF}(q)^d$. Let us call these the "McFarland type" constructions. It might be possible to use less trivial objects in the vector space $\mathrm{GF}(q)^d$ to find new constructions of (divisible) difference sets. In my opinion, it is not too interesting to find just new series of divisible difference sets (there will be plenty) but to restrict to difference sets with some additional properties (for instance almost difference sets, difference sets with certain multipliers, semiregular examples). We did not investigate difference sets with multiplier -1 here. They have been investigated quite extensively. For the (classical) difference set case, we refer to Jungnickel [100] and the references quoted there. Relative difference sets with multiplier -1 have been studied by Ma [125], for the divisible case, the reader is referred to Arasu, Jungnickel and Pott [12] and Leung, Ma and Tan [119].

2.4   Non-existence results

In the last section, we have seen several constructions of divisible difference sets. If we made a table of admissible parameters of divisible difference sets we could fill already many entries with a "yes", this difference set exists. But there are still many entries where we do not know a construction. In this case, one can try to construct an example or can try to prove that no difference set can exist. In this section, we will summarize several non-existence results for divisible difference sets. Some statements use the existence of multipliers, see Theorem 1.3.5. However, in many cases, the multiplier theorems are not strong enough to prove that certain integers have to be numerical multipliers. Therefore, multiplier free non-existence results are of particular interest. A classical result in this direction is due to Mann [131]. It can be generalized easily to divisible difference sets. The following version is a slight generalization of the Mann test in [14].

Theorem 2.4.1 (Mann test; Arasu, Jungnickel, Pott [14]) *Let R be a divisible $(m, n, k, \lambda_1, \lambda_2)$-difference set in G relative to N. Let $\chi : G \to \mathbf{C}^*$ be a non-principal character of order w. Moreover, let t be a χ-multiplier and p a prime not dividing w which satisfies*

$$tp^f \equiv -1 \ mod \ w$$

for some integer f. Then the following conclusions hold:

(a) *If the kernel of χ does not contain N, then p does not divide the square-free part of $k - \lambda_1$ provided that $k - \lambda_1 \neq 0$.*

(b) *If the kernel of χ contains N, then p does not divide the square-free part of $k^2 - \lambda_2 mn$ provided that $k^2 - \lambda_2 mn \neq 0$.*

Proof. (a) Let us assume that p divides $k - \lambda_1$, otherwise there is nothing to prove. Let p^i be the largest power of p dividing $k - \lambda_1$. Note that $\chi(R)\overline{\chi(R)} = k - \lambda_1$. Let P be the greatest common (ideal) divisor of $(p)^i$ and $(\chi(R))$. Then $\overline{P}|(\overline{\chi(R)})$ and $P\overline{P} = (p)^i$: There is an ideal Q dividing $(p)^i$ and $(\overline{\chi(R)})$ with $PQ = (p)^i$ and therefore $\overline{P}|Q$ and $P|\overline{Q}$. Since $\overline{Q}|(\chi(R))$ and $\overline{Q}|(p)^i$, we obtain $\overline{P} = Q$. But P as well as Q are fixed under the Galois automorphism $\zeta_w \mapsto \zeta_w^t$ (since t is a χ-multiplier) and $\zeta_w \mapsto \zeta_w^{p^f}$ (in view of Result 1.2.7). Therefore, P is fixed under $\zeta_w^{tp^f}$ which is assumed to be "complex conjugation". We obtain $P = \overline{P} = Q$ and $P^2 = (p)^i$. This shows that i is an even number since the prime ideal divisors of (p) in $\mathbf{Z}[\zeta_w]$ are distinct (Result 1.2.7). The proof of (b) is quite similar with the only modification that $\chi(R)\overline{\chi(R)} = k^2 - \lambda_2 mn$ since χ is principal on N. □

If $w = 2$, we can say more since in this case $\mathbf{Z}[\zeta_2] = \mathbf{Z}$.

Proposition 2.4.2 *Let R be a divisible $(m, n, k, \lambda_1, \lambda_2)$-difference set in G relative to N. Let χ be a character of order 2. Then the following holds:*

(a) *If the kernel of χ does not contain N, then $k - \lambda_1$ is a square.*

(b) *If the kernel of χ contains N, then $k^2 - \lambda_2 mn$ is a square.* □

Proposition 2.4.3 *There is no splitting abelian relative $(n, n, n, 1)$-difference set of odd order n with $n \leq 100$, $n \neq 39, 55, 63, 93$ where n is not a prime power.*

Proof. The following table shows the values w, p and f which we have to choose in Theorem 2.4.1:

n	p	w	p^f
15	3	5	3^2
21	3	7	3^3
33	11	3	11^1
35	7	5	7^2
45	5	3	5^1
51	17	2	17^1
57	3	19	3^7
65	13	5	13^2
69	23	3	23^1
75	3	5	3^2
77	7	11	11^3
85	17	5	17^2
87	29	3	29^1
95	19	5	19^1
99	11	3	11^1

□

It is possible to show the nonexistence of the relative difference sets with $n = 39$ and $n = 93$, see Theorem 5.4.2. The cases 55 and 63 are still open! Note that examples of these difference sets always exist if n is a prime power (Theorems 2.2.9 and 2.2.10). We have considered only the splitting case since it is, from a geometric point of view, the most interesting case (Section 5.4). We have restricted ourselves to the odd case since it is known that n has to be a power of 2 provided that n is even, see Result 5.4.1.

Example 2.4.4 There is no divisible $(7, 41, 139, 129, 57)$-difference set. We apply Theorem 2.4.1 with $t = 1$, $p = 2$ and $w = 41$. Note that the subgroup of order 7 is a normal subgroup with abelian quotient group, therefore there is a character of order 41. We have $2^{10} \equiv -1 \bmod 41$, therefore no such divisible difference set can exist since 2 divides the square-free part of $k - \lambda_1 = 10$. However, a symmetric divisible design with these parameters exist by Theorem 2.1 in Arasu, Haemers, Jungnickel and Pott [9].

Example 2.4.5 There is no abelian splitting divisible difference set with parameters $(14, 2, 11, 6, 4)$ in a group G. The (putative) difference set has $k - \lambda_1 = 5$ and $k^2 - \lambda_2 mn = 9$. We can use Theorem 2.4.1 with $w = 2$, $t = 1$ and $p = 5$; we have $p \equiv -1 \bmod 2$. not use this argument if the Sylow 2-subgroup of G is cyclic (i.e. the difference set is non-splitting) since in that case each character of order 2 has N in its kernel. Since $k^2 - \lambda_2 mn$ is a square, the necessary condition of Theorem 2.4.1 is satisfied. This motivates the following extension of the Mann test:

Theorem 2.4.6 (Arasu, Jungnickel, Pott [14]) *Let R be an $(m, n, k, \lambda_1, \lambda_2)$-divisible difference set in a group G relative to a subgroup N. Let U be a normal subgroup of G such that G/U is an abelian group of exponent w. Moreover, let t be a χ-multiplier for <u>all</u> characters whose kernel contains U. Let p be a prime which does not divide w and which satisfies*

$$tp^f \equiv -1 \; mod \; w$$

for some integer f. We define the integer i to be the largest exponent such that p^i divides $k - \lambda_1$ if N is not contained in U, or the largest exponent such that p^i divides $k^2 - \lambda_2 mn$ if N is a subgroup of U, provided that the respective numbers are not 0. Then the integer i is even and $p^{i/2} \leq |U|$.

Proof. The statement that i is even (say $i = 2j$) is already contained in Theorem 2.4.1. We denote the canonical epimorphism from G onto G/U and its extension from $\mathbf{Z}[G]$ onto $\mathbf{Z}[G/U]$ by ρ and put $\rho(G) =: H$. Let us begin with the case that U does not contain the subgroup N. We will show that the coefficients of

$$\rho(R) := \sum_{h \in H} t_h h$$

are constant modulo p^j on cosets of $\rho(N)$. The inversion formula shows

$$|H| t_h = \sum_{\chi \in H^*} \chi(\rho(R)) \chi(h^{-1}).$$

Since $|H|$ and p are relatively prime, it is enough to show that $(t_h - t_{hg}) \equiv 0 \; mod \; p^j$ for $g \in \rho(N)$. Let χ be a character of H which is non-principal on $\rho(N)$. The proof of Theorem 2.4.1 tells us that $\chi(\rho(R)) \equiv 0 \; mod \; p^j$ holds for all characters $\chi \in H^*$ which are non-principal on $\rho(N)$ (note that we assume that t is a χ-multiplier for all characters whose kernel contains U and not only for characters of order w). We obtain

$$t_h - t_{hg} \equiv \sum_{\chi \in H^*} \chi(\rho(R))(\chi(h^{-1}) - \chi(h^{-1}g^{-1})) \equiv 0 \; mod \; p^j.$$

The t_h's are nonnegative and bounded by $|U|$. If $|U| < p^j$, then the coefficients of $\rho(R)$ would be constant on cosets of $\rho(N)$, say $\rho(R) = \rho(N)A$. But in that case $\chi(\rho(R)) = 0$ for characters non-principal on $\rho(N)$ (note that such characters exist since N is not contained in U). This contradicts $k - \lambda_1 \neq 0$ since characters of H non-principal on $\rho(N)$ can be lifted to characters of G non-principal on N.

The case that N is a subgroup of U is easier since in this case there is no exceptional subgroup $\rho(N)$: We can conclude, as above, that all the coefficients in $\rho(R)$ are congruent modulo p^j, and we reach a contradiction if $|U| < p^j$. $\square$

Example 2.4.7 We can rule out the existence of a non-splitting abelian $(14, 2, 11, 6, 4)$-difference set relative to N using Theorem 2.4.6: We choose a character of order 14 (whose kernel is N). Using the prime $p = 3$, we get $3^3 \equiv -1 \; mod \; 14$, $i = 2$, and $U = N$, hence $3 \leq |U| = 2$, a contradiction.

Example 2.4.8 There is no splitting abelian divisible $(7, 56, 179, 170, 67)$-difference set relative to N. Here we choose a character whose order divides 28, and whose kernel is a subgroup of order 2 in N (this is possible since we are in the splitting case). We apply Theorem 2.4.6 with $t = 1$, $p = 3$, and $3^3 \equiv -1 \bmod 28$. With $j = 1$, we reach a contradiction $3 \leq 2$. A symmetric divisible design with these parameters exist: We can apply Theorem 2.1 in [9] with $s = 2$ and a symmetric $(56, 11, 2)$-design.

The next example is of a different flavour. We cannot rule out the existence of that divisible difference set directly. But it is possible to handle it with a little ad hoc argument. Such ad hoc arguments can be used quite often to solve difference set problems.

Example 2.4.9 There is no abelian divisible difference set R with parameters $(3, 112, 149, 124, 37)$ in G. If G is not cyclic, we can use Theorem 2.4.1: We choose a subgroup of order 4 such that the exponent of the quotient group is at most 42. Then $5^3 \equiv -1 \bmod 42$ yielding the contradiction $5 < 4$. If G is cyclic, we apply a character of order 3. We can write $\chi(R) = a\zeta + b\zeta^2$, where ζ is a complex 3rd root of unity. We compute

$$\chi(R)\overline{\chi(R)} = a^2 + b^2 - ab = k^2 - \lambda_2 mn = 9769$$

using $1 + \zeta + \zeta^2 = 0$ and the fact that the kernel of χ must contain N. Now it is easy to check (with a computer) that this equation has no integral solution.

The two Theorems 2.4.1 and 2.4.6 are a very important tool to prove the non-existence of putative difference sets. In case that the theorems cannot be applied directly, one can use the following strategy. Begin with the putative (divisible) difference set D, and project it onto some smaller group H. Try to prove that the hypothetical image of D cannot exist. In order to do this, it is useful to choose the group H such that a prime p is self-conjugate modulo the exponent of H. Many arguments of this type are discussed in Lander's book [116]. At this point, we want to prove Turyn's famous exponent bound for Hadamard difference sets.

Theorem 2.4.10 (Turyn [167]) *Suppose that there exists an abelian Hadamard difference set D with parameters $(4u^2, 2u^2 - u, u^2 - u; u^2)$ in G. Then the following holds:*

(a) *If $u = 2^a$, then $exp(G) \leq 2^{a+2}$.*

(b) *If $u = p^a$ $(p \neq 2$, p prime) and T denotes a Sylow p-subgroup of G, then $exp(T) \leq p^a$.*

Proof. (a)   Let 2^b be the exponent of G. We can show that the image of D in a cyclic group of order 2^b cannot exist if $b > a + 2$: We project G onto a quotient group H which is cyclic of order 2^b. This projection is denoted by ρ.

The coefficients of $\rho(D)$ are bounded by 2^{2a+2-b}. On the other hand, we can write $\rho(D)$ in the form

$$\rho(D) = 2^a X + PY$$

(see Corollary 1.2.14) where P is the unique subgroup of order 2 in H. If $2^a > 2^{2a+2-b}$ (equivalently $b > a + 2$), then the coefficients of $\rho(D)$ would be constant on cosets of P. But this is impossible since

$$\rho(D)\rho(D)^{(-1)} = 2^{2a} + 2^{2a+2-b}(2^{2a} - 2^a)H.$$

(b) In the second case, it is not necessarily true that the prime p is self-conjugate modulo the exponent of G: If the Sylow 2-subgroup of G is cyclic of order 4 and $p \equiv 1 \bmod 4$, p is not self-conjugate. But p is always self-conjugate modulo 2 times a power of p. Hence we project onto a cyclic group H of order $2p^b$ where p^b is the exponent of T. The image $\rho(D)$ can be written in the form

$$\frac{\rho(D)}{2} = p^a X + PY$$

where P is again the subgroup of H of order p. The coefficients of $\rho(D)$ are bounded by $2p^{2a-b}$ which shows (as above) $p^a \leq 2p^{2a-b}$. $\qquad \square$

It is important to mention that this exponent bound is not only necessary but also <u>sufficient</u> for the existence of a Hadamard difference set if u is a power of 2, see Result 2.1.7. If u is a power of 3, it is known that the exponent bound is not sufficient for the existence of a difference set (Example 2.4.13). If u is not a power of 2, several non-existence results are known. We refer the reader to Davis and Jedwab [53].

We can strengthen the exponent bound in case that u is even (but not necessarily a power of 2):

Theorem 2.4.11 (Turyn [167]) *The Sylow 2-subgroup S of an abelian Hadamard difference set D of even order u^2 cannot be cyclic.*

Proof. Let 2^t be the largest power of 2 which divides u. Let D' be the image of D under the projection on $\mathbf{Z}[S]$. Since 2 is self-conjugate modulo $|S| = 2^{2t+2}$, we have $\chi(D') \equiv 0 \bmod 2^t$ for all characters χ of S. Let ρ denote the canonical epimorphism $S \to S/S'$ where S' is the subgroup of S of order 2^t. Proposition 1.2.12 shows that the coefficients of $\rho(D')$ are divisible by 2^t. We put $D'' = \rho(D')/2^t = \sum d_i$ and $u_1 = u/2^t$. We obtain

$$D'' D''^{(-1)} = u_1^2 + (2^t u_1^4 - u_1^3)S/S' = u_1^2 + u_1^3(u-1)S/S'.$$

Consider the coefficient of the identity modulo 2. The coefficient is

$$u_1^2 + u_1^3(u-1) = \sum (d_i)^2 \equiv \left(\sum d_i\right)^2 = u_1^2 + u_1^3(u-1)2^{t+2} \bmod 2.$$

This is only possible if $u = 1$. $\qquad \square$

Let us look at the proof of Theorem 2.4.10 more closely. Assume that $p \neq 2$
and that p is self-conjugate modulo the exponent of G. Suppose that p^{2a} is the
order of the Sylow p-subgroup of G with exponent p^a. Then we can write

$$\rho(D) = p^a X + PY \tag{2.19}$$

in $\mathbf{Z}[\rho(G)]$ where X has only 0/1-coefficients (hence X can be viewed as a subset
of $\rho(G)$), and X and Y have no common coefficient different from 0. This follows
from the fact that the coefficients of $\rho(D)$ are bounded by p^a. We may assume
that no coset of P is contained in X. The question arises whether we can say
more about solutions of (2.19) by a more sophisticated analysis. This is indeed
true using a trick due to Arasu, Davis and Jedwab [5]. We assume that there
exists a Hadamard difference set of order u^2 in an abelian group $G \cong H \times K \times \mathbf{Z}_{p^a}$
where p^a is the largest power of p dividing u and $|K| = p^a$. Moreover, suppose
that p is self-conjugate modulo the exponent of H. Let ρ denote the canonical
epimorphism from G onto G/K, and let P be the unique subgroup of order p
in $\rho(G)$. We can write $\rho(D)$ as in (2.19), and we assume that no coset of P
is contained in X. Let w denote the number of cosets of P which meet X (by
assumption, they are not contained in X). In other words, w is the number of
cosets K' of $K \times P$ such that a coset of K in K' is contained in D, but K' is
not completely contained in D. Note that $K' \cap D$ is the union of cosets of K,
otherwise the coefficients of $\rho(D)$ are not constant modulo p^a on cosets of P.
    The element $\rho(D)$ satisfies the equation

$$\rho(D)\rho(D)^{(-1)} = u^2 + (u^2 - u)p^a \rho(G). \tag{2.20}$$

Using this equation, it is possible to compute a lower bound for w. We write

$$Y := \sum_{g \in T} y_g g$$

where T denotes a system of distinct coset representatives of P in $\rho(G)$. We
define

$$w_g := |X \cap Pg|$$

and consider the coefficient of the identity in (2.20). We obtain

$$p^{2a} \sum w_g + p \sum (y_g)^2 = u^2 + (u^2 - u)p^a. \tag{2.21}$$

Now we are using a further projection μ from $\rho(G)$ onto $\rho(G)/P$. Then $\mu(\rho(D))$
satisfies

$$\mu(\rho(D))\mu(\rho(D))^{-1} = u^2 + (u^2 - u)p^{a+1}\mu(\rho(G)).$$

Again, we compute the coefficient of the identity. The contribution of $\mu(PY)$ to
this coefficient is

$$p^2 \sum (y_g)^2,$$

the contribution of $p^a \mu(X)$ is

$$p^{2a} \sum (w_g)^2,$$

which gives the equation

$$p^{2a} \sum (w_g)^2 + p^2 \sum (y_g)^2 = u^2 + (u^2 - u)p^{a+1}. \tag{2.22}$$

Combining (2.21) and (2.22), we get

$$\sum (pw_g - w_g^2) = \frac{u^2(p-1)}{p^{2a}}$$

and

$$0 \le pw_g - w_g^2 \le \frac{p(p+1)}{2} - \frac{(p+1)^2}{4} = \frac{p^2 - 1}{4}$$

(note that p is odd). This shows

$$w \cdot \frac{p^2 - 1}{4} \ge \frac{u^2(p-1)}{p^{2a}},$$

hence

$$w \ge \frac{4u^2}{p^{2a}(p+1)}. \tag{2.23}$$

We are now ready to prove the following Theorem:

Theorem 2.4.12 (Arasu, Davis, Jedwab [5]) *Let D be an abelian Hadamard difference set of order u^2 in G where p^a is the highest power of p (p odd) which divides u. Suppose that p is self-conjugate modulo the exponent of G. If the Sylow p-subgroup of G is isomorphic to $K \times \mathbf{Z}_{p^a}$ (i.e. the exponent is p^a), then K has to be cyclic.*

Proof. We denote the rank of K by r, i.e.,

$$K := \langle k_1, \ldots, k_r \rangle$$

is generated by $k_1, \ldots, k_r$. Moreover, let z be a generator of the (unique) cyclic subgroup P of $\mathbf{Z}_{p^a} < G$ with $|P| = p$. Let $i = (i_1, \ldots, i_r)$ be a vector of length r with $0 \le i_1, \ldots, i_r \le p - 1$. The groups

$$H_i := \langle k_1 z^{i_1}, \ldots, k_r z^{p_r} \rangle$$

are isomorphic to K, and they intersect K in exactly p^{a-1} elements (unless $i = (0, \ldots, 0)$ when $K \cong H_0$). Moreover, the groups H_i are complements of $\mathbf{Z}_{p^a}$ in the Sylow p-subgroup of G. They are all contained in $K \times \langle z \rangle$, and the intersection of H_i and H_j ($i \ne j$) is contained in K. Let μ_i denote the canonical epimorphism $G \to G/H_i$. The unique subgroup of order p in $\mu_i(G)$ is generated by $\mu_i(z)$. A coefficient of $\mu_i(D)$ is p^a if and only if a coset of H_i is contained in D. There are two cases: Either a coset of $K \times \langle z \rangle = H_i \times \langle z \rangle$ (say K') is contained in D or this is not the case. We know that the latter case occurs at least $w \ge 4u^2/p^{2a}(p+1)$ times, in which case K' gives rise to p coefficients a_{gz^i}, $i = 0, \ldots, p-1$ which are multiples of p^{a-1} (but $\ne p^a$) under the projection ρ

onto G/K. Is it possible that two cosets H_i' and H_j' (contained in D) of H_i and H_j give rise to the same p coefficients under the projection onto G/K? If this is true, then H_i' and H_j' are in the same coset K' of $K \times P$, and K' meets each coset of H_i contained in K' at least once (since H_j' is in D). But since K' is the union of cosets of H_i, this implies $K' \subset D$, but we have assumed that K' is not contained in D.

Combining this with the bound on w in 2.23, we obtain the following: There are at least $4u^2/(p^{2a}(p+1))$ cosets of P in $\rho(G)$ on which the coefficients of $\rho(D)$ are 0 and p^a. Moreover, there are at least $(4u^2/(p^{2a}(p+1)))p(p^r - 1)$ further coefficients which are multiples of p^{a-1} (but different from p^a). This is enough to prove $r = 1$: In order to do this, we replace D by $D' := 2D - G$, see also Section 6.1. The nice thing about this transformation is that every projection $\tau(D')$ satisfies $\tau(D')\tau(D'^{(-1)}) = 4u^2$. We have

$$\rho(D') = 2\rho(D) - p^a G/K.$$

All we have said about $\rho(D)$ translates into a statement about $\rho(D')$. In particular, coefficients 0 and p^a in $\rho(D)$ become coefficients of absolute value p^a. Coefficients which are multiples of p^{a-1} but are different from p^a become (in absolute value) multiples of p^{a-1} different from p^a. Computing the coefficient of the identity in $\rho(D')\rho(D'^{(-1)})$, we get

$$\frac{4u^2}{p^{2a}(p+1)} \left(p \cdot p^{2a} + p(p^r - 1)p^{2a-2} \right) \leq 4u^2$$

which shows $r = 1$.   □

Example 2.4.13 There is no abelian Hadamard difference set with parameters $(364, 153, 72; 81)$ in $H \times \mathbf{Z}_3 \times \mathbf{Z}_3 \times \mathbf{Z}_9$ where H is any of the two groups of order 4. However, the group $H \times \mathbf{Z}_9 \times \mathbf{Z}_9$ has a difference set, see Result 2.1.7. This example shows that the exponent bound in Theorem 2.4.10 is not sufficient for the existence of a Hadamard difference set if the order is a power of 3.

There are quite a few more necessary conditions on the existence of Hadamard difference sets which are not stated in this monograph. The deepest result in this area is the following theorem due to McFarland:

Result 2.4.14 (McFarland [135]) *An abelian Hadamard difference set of order $4p^2$ cannot exist if p is a prime different from 2 and 3.*

It should be mentioned that the case $p \equiv 3 \mod 4$ is comparitively easy (see Mann and McFarland [132]). The difficult stuff is the case $p \equiv 1 \mod 4$. This result provided some evidence on the conjecture that Hadamard difference sets of order u^2 exist if and only if the only prime divisors of u are 2 and 3, a conjecture that has been disproved spectacularly by Xia [177], see Result 2.1.7 and also Xiang [178] for a more transparent proof.

Using the non-existence theory developed so far together with McFarland's Result 2.4.14, there are eight groups left where the existence of an abelian Hadamard difference set of order u^2 with $u \leq 20$ is still undecided. These cases are (see Davis and Jedwab [53])

$$\mathbf{Z}_2^2 \times \mathbf{Z}_4 \times \mathbf{Z}_5^2, \quad \mathbf{Z}_2 \times \mathbf{Z}_8 \times \mathbf{Z}_5^2, \quad \mathbf{Z}_4^2 \times \mathbf{Z}_5^2, \quad \mathbf{Z}_2^2 \times \mathbf{Z}_{16} \times \mathbf{Z}_9,$$
$$\mathbf{Z}_4 \times \mathbf{Z}_{16} \times \mathbf{Z}_9, \quad \mathbf{Z}_8^2 \times \mathbf{Z}_9, \quad \mathbf{Z}_4 \times \mathbf{Z}_3^2 \times \mathbf{Z}_5^2, \quad \mathbf{Z}_2 \times \mathbf{Z}_8 \times \mathbf{Z}_3^2 \times \mathbf{Z}_9.$$

The first three cases above ($u = 10$) have been ruled out recently with a computer by Pollatsek [140]. It is not the intention of this monograph to develop the non-existence theory about difference sets in full generality but to summarize the main techniques. The technique to prove Theorem 2.4.12 is, in my opinion, rather interesting and does appear also in connection with relative semiregular difference sets.

We are now going to describe another "trick" to solve difference set problems. Many theorems about (divisible) difference sets in G require the existence of a prime which is self-conjugate modulo the exponent of some quotient group of G. Recently, some effort has been made to weaken this assumption. The technique which we will use has been developed to obtain necessary conditions on the existence of Hadamard difference sets, see Chan [43]. Here we state the following theorem whose proof has been inspired by Chan:

Theorem 2.4.15 (Arasu, Pott [19]) *Let p and q be distinct primes such that the order f of q modulo p is odd. Let ζ_p be a primitive p-th root of unity in $\mathbf{C}$. We define $p = ef + 1$. If $\gamma \in \mathbf{Z}[\zeta_p]$ satisfies $\gamma\bar{\gamma} = n$ and the Galois automorphism determined by $\zeta_p \mapsto \zeta_p^q$ fixes the prime ideal divisors of n in $\mathbf{Z}[\zeta_p]$, then there exists a suitable integer y such that*

$$\gamma\zeta_p^y = \sum_{i=1}^{e} b_i \gamma_i, \quad b_i \in \mathbf{Z}$$

where

$$\gamma_i := \sum_{j=0}^{f-1} (\zeta_p)^{iq^j}$$

and the b_i satisfy

$$n = \frac{p \sum (b_i)^2 - f(\sum b_i)^2}{e}, \quad \sum b_i \leq e\sqrt{n}. \tag{2.24}$$

Proof. We write γ in the form $\gamma = \sum_{i=1}^{p-1} c_i \zeta_p^i$. This representation is unique since the ζ_p^i, $i = 1, \ldots, p-1$, form an integral basis, see Result 1.2.3. By assumption, the prime ideal divisors of n are fixed by the Galois automorphism σ defined by $\zeta_p \mapsto \zeta_p^q$. Hence the ideal generated by γ is the same as the ideal generated by $\sigma(\gamma)$ (in $\mathbf{Z}[\zeta_p]$). This shows $\sigma(\gamma) = \mu\gamma$ for a suitable unit μ in the ring $\mathbf{Z}[\zeta_p]$. The absolute value of γ and all its algebraic conjugates is $\sqrt{n}$, therefore $|\mu| = 1$, and the same holds for all algebraic conjugates of μ. This proves that μ is a root of unity, say $\pm(\zeta_p)^j$ (see Result 1.2.8). First, we will show that μ is $+(\zeta_p)^j$. We consider the product

$$\gamma\,\sigma(\gamma)\,\sigma^2(\gamma)\,\ldots\,\sigma^{f-1}(\gamma) = \mu\,\sigma(\mu)\,\sigma^2(\mu)\,\ldots\,\sigma^{f-1}(\mu)\,\gamma\,\sigma(\gamma)\,\sigma^2(\gamma)\,\ldots\,\sigma^{f-1}(\gamma).$$

If $\mu = -(\zeta_p)^j$ then $1 = -(\zeta_p)^x$ for some $x \neq 0$ which is impossible (note that f is odd). Hence we have $\mu = \zeta_p{}^j$, therefore $\sigma(\gamma) = \gamma\zeta_p{}^j$ and $\sigma(\gamma\zeta_p{}^y) = \gamma\zeta_p{}^{j+yq}$ for integers j and y. We choose $y \in \{0, \ldots, p-1\}$ such that $y \equiv j + yq \bmod p$, and then we may assume that γ is $\underline{\text{fixed}}$ under σ if we replace γ by $\gamma\zeta_p{}^y$. Observe that σ fixes the integral basis $\zeta_p, \zeta_p{}^2, \ldots, \zeta_p{}^{p-1}$ setwise, hence $\gamma = \sum_{i=1}^{e} b_i\gamma_i$ with $b_i \in \mathbf{Z}$.

Now we have

$$\left(\sum b_i\gamma_i\right)\left(\sum b_i\overline{\gamma_i}\right) = -n(\zeta_p + \ldots + \zeta_p{}^{p-1}).$$

We count the sum of the coefficients of the $\zeta_p{}^i$ $(i = 1, \ldots, p-1)$ to obtain

$$-n(p-1) = f^2\left(\sum b_i\right)^2 - pf\sum(b_i)^2$$

which proves the equality in (2.24). The inequality in (2.24) can be proved using the Cauchy-Schwarz inequality: We have $\left(\sum b_i\right)^2 \leq e\sum(b_i)^2$, therefore,

$$n(p-1) = pf\sum(b_i)^2 - f^2\left(\sum b_i\right)^2 \geq (pf/e - f^2)\left(\sum b_i\right)^2$$

and

$$e^2 n \geq (p - ef)\left(\sum b_i\right)^2 = \left(\sum b_i\right)^2.$$

$\square$

Corollary 2.4.16 *Let R be a divisible difference set in G relative to N with parameters $(m, n, k, \lambda_1, \lambda_2)$. Assume that G admits a character of order $p = ef + 1$ with $\chi(R)\chi(R^{(-1)}) = r$, where*

$$r = \begin{cases} k^2 - \lambda_2 mn & \text{if } \chi|N = \chi_0 \\ k - \lambda_1 & \text{otherwise.} \end{cases}$$

If $r = s^2 t$, t not a square, and if q is a χ-multiplier of odd order f, then $p \leq e^2 t$ provided that the prime divisors of s are self-conjugate modulo p.

Proof. We write $\chi(R) = \sum_{i=1}^{e} b_i\gamma_i$ and obtain

$$r = \frac{p\sum(b_i)^2 - f(\sum b_i)^2}{e}, \qquad \sum b_i \leq es\sqrt{t}.$$

Note that the prime ideal divisors of r are fixed by the Galois automorphism $\zeta_p \mapsto \zeta_p{}^q$ since q is a χ-multiplier. Since the prime divisors of s are self-conjugate modulo p, we have $\chi(R) \equiv 0 \bmod s$, and thus we can divide each of the b_i by s (note that the $\zeta_p{}^i$ form an integral basis). We obtain (with $a_i := b_i/s$)

$$
\begin{aligned}
te &= p\sum(a_i)^2 - f(\sum a_i)^2, \quad (\sum a_i)^2 \le te^2 \\
te^2 &= ep\sum(a_i)^2 - (p-1)(\sum a_i)^2, \\
te^2 &\equiv (\sum a_i)^2 \bmod p.
\end{aligned}
$$

If $p > te^2 \ge (\sum a_i)^2$, we get a contradiction since te^2 is not an integer square. $\square$

We will now give some applications of Theorem 2.4.15 and its corollary.

Proposition 2.4.17 (Arasu, Pott [19]) *Abelian relative difference sets with parameters $(n+1, n-1, n, 1)$ do not exist if $n = 5s^4$ with $s \equiv 3 \bmod 101$ or $n = 3s^6$ with $s \equiv 5 \bmod 23$ (in both cases, we assume that s is a prime).*

Proof. In the first case, we apply Corollary 2.4.16 with $p = 101$, i.e., we take a character χ of order 101 non-principal on the forbidden subgroup N. Note that 101 is a divisor of $n-1$. The order of 3 modulo 101 is 100, therefore s is self-conjugate modulo p. Moreover, 5 is a χ-multiplier of order 25 modulo 101, and thus we have $e = 4$ and $t = 5$ using the notation above. This yields the contradiction $101 \le 16 \cdot 5$. Note that 5 is a χ-multiplier since the automorphism $\zeta_{101} \mapsto \zeta_{101}{}^5$ fixes prime ideal divisors of 5 (see Result 1.2.7) and this automorphism is a power of the automorphism $\zeta_{101} \mapsto \zeta_{101}{}^3$ which fixes prime ideal divisors of n (see Result 1.2.7 again).

In the second case, we choose $p = 23$ (which divides $n-1$). Note that s is self-conjugate modulo 23 (since the order of 5 modulo 23 is 22). The order of the χ-multiplier 3 is 11, hence $e = 2$ and $t = 3$. Our corollary yields $23 \le 4 \cdot 3$, a contradiction. $\square$

Two remarks are in order concerning this proposition. The second series can be also ruled out using a result of Yamamoto [180]. However, the use of Corollary 2.4.16 seems to be much easier than solving the diophantine equation that appear in Yamamoto's theorem. Second, the numbers 3 and 5 are not only χ-multipliers but are actually **multipliers** of the difference sets (Theorem 1.3.5). In the next example, I do not know a multiplier theorem that shows that 5 would be a multiplier.

Proposition 2.4.18 (Arasu, Pott [19]) *Abelian $(4n-1, 2n-1, n-1; n)$-difference sets do not exist if $n = 5s^2$ where s is a prime congruent 13 modulo 109.*

Proof. We choose $p = 109$ (which divides $4n-1$). The prime s is self-conjugate modulo p since the order of 13 modulo 109 is 108. Moreover, 5 is a χ-multiplier: The Galois automorphism $\zeta_{109} \mapsto \zeta_{109}{}^5$ fixes the prime ideal divisors of 5 and since it is also a power of $\zeta_{109} \mapsto \zeta_{109}{}^5$ (which fixes prime ideal divisors of s), it fixes prime ideal divisors of s. With $e = 4$ and $t = 5$, we reach a contradiction $109 \leq 16 \cdot 5$.   □

Chapter 3

Difference sets with classical parameters

We say that an (m, k, λ)-difference set has the classical parameters if

$$(m, k, \lambda) = \left(\frac{q^{d+1} - 1}{q - 1}, \frac{q^d - 1}{q - 1}, \frac{q^{d-1} - 1}{q - 1} \right) \tag{3.1}$$

or if it has the complementary parameters

$$\left(\frac{q^{d+1} - 1}{q - 1}, q^d, q^d - q^{d-1} \right) \tag{3.2}$$

A relative (m, n, k, λ)-difference set has the classical parameters if

$$(m, k, n\lambda) = \left(\frac{q^d - 1}{q - 1}, q^{d-1}, q^{d-2} \right), \tag{3.3}$$

in other words, the relative difference set projects onto a difference set with the (classical) parameters (3.2) of the complements of the Singer difference sets. We have noticed already that all <u>known</u> non-trivial relative difference sets with $n \neq 2$ have the classical parameters. Parametrically, most of them have the parameters of the classical affine difference sets (see (2.9)) or projections of them. There are two exceptions: One is the small sporadic $(21, 6, 16, 2)$-difference set (Example 2.2.14), the other series will be constructed in Section 3.3, see also Theorem 2.2.13. But first, we will study the (classical) Singer difference sets and the GMW-difference sets in the first two sections of this chapter.

3.1 Singer difference sets

Recall that the powers of a primitive element α of $GF(q^{d+1})$ with trace$(\alpha^i) = 0$ considered modulo $(q^{d+1} - 1)/(q - 1)$ form a classical Singer difference set. We will first show that the only multipliers of the classical Singer difference sets are the powers of p.

Proposition 3.1.1 *Let D be a classical Singer difference set with parameters (3.1). If t is a multiplier of D, then t has to be a power of p (modulo $(q^{d+1} - 1)/(q-1)$) where $q = p^e$ is a power of the prime p.*

Proof. We know already that the powers of p are multipliers of D, hence the multiplier group has order at least $e(d+1)$. To prove that these are the only multipliers, we will use Theorem 1.3.1. The full automorphism group of the point-hyperplane design of $PG(d, q)$ is $P\Gamma L(d+1, q)$. This is a classical result from projective geometry, see [30], for instance. This group is the semi-direct product of the "field automorphisms" of $GF(q)$ (there are e such automorphisms) and $PGL(d+1, q)$. In order to show that the multiplier group M has exactly $e(d+1)$ elements, it is obviously enough to show $|M \cap PGL(d+1, q)| = d+1$. For this purpose, let $\mathbf{W}$ be an element of order $q^{d+1} - 1$ in $GL(d+1, q)$, hence the powers of $\mathbf{W}$ together with the zero-matrix are the field $GF(q^{d+1})$. The multiplicative group of $GF(q^{d+1})$ modulo $\langle \mathbf{W}^{(q^{d+1}-1)/(q-1)} \rangle$ is our Singer group. (For a thorough discussion of the representation of a finite field as a matrix field, we refer the reader to Jungnickel [101].) The minimum polynomial of $\mathbf{W}$ considered as an element in $GL(d+1, q)$ is the same as the minimum polynomial of $\mathbf{W}$ considered as an element in the extension field $GF(q^{d+1})$ of $GF(q)$. If s is a multiplier contained in $PGL(d+1, q)$, then we have $\mathbf{W}^s = \lambda \mathbf{A}^{-1} \mathbf{W} \mathbf{A}$ for an element $\mathbf{A} \in GL(d+1, q)$ and $\lambda \in GF(q)^*$. In particular, $\lambda^{-1} \mathbf{W}^s = \mathbf{W}^{j(q^{d+1}-1)/(q-1)+s}$ (for a suitable integer $0 \le j \le q-1$) and $\mathbf{W}$ have the same minimum polynomial. But the only powers of $\mathbf{W}$ with the same minimum polynomial as $\mathbf{W}$ are $\mathbf{W}^q, \ldots, \mathbf{W}^{q^d}$ which shows $|M \cap PGL(d+1, q)| = d+1$.   □

Corollary 3.1.2 *The only multipliers of the classical affine difference sets with parameters (2.9) are the powers of p (where $q = p^e$).*

Proof. Since the classical affine difference sets project onto the classical Singer difference sets (more precisely, their complements), and since these have only the powers of p as multipliers, the corollary is proved. This argument does not work if $d = 2$, since in this case the underlying difference set is trivial, hence it admits all integers relatively prime to $q+1$ as multipliers: In this case, one simply has to repeat the proof of Proposition 3.1.1.   □

Now let us investigate the problem how to generate Singer difference sets. In general, it seems not easy to compute the trace of all field elements. But it is enough to know the trace of only a few elements: We consider the field extension $GF(q^{d+1})/GF(q)$ with primitive element α. We assume that the values

$$a_i = \text{trace}(\alpha^i)$$

are known for $0 \le i \le d$. If

$$x^{d+1} - c_d x^d - c_{d-1} x^{d-1} - \ldots - c_0$$

is the minimum polynomial of α over $GF(q)$ (hence a so called primitive polynomial), we obtain

$$a_{d+1} = c_d a_d + \ldots + c_0 a_0$$

and, more generally,

$$a_{d+i} = c_d a_{d+i-1} + \ldots + c_0 a_{i-1}.$$

We say that the sequence (a_i) satisfies a **linear recurrence relation of degree** $d+1$. In view of the arguments above, it is enough to know the trace function only for some powers of α in order to determine the trace function for all the elements quite easily using the linear recurrence relation.

If we just want to find <u>one</u> Singer difference set (not necessarily the trace 0 hyperplane), the problem is even easier. Assume there are two elements α^i and α^j such that

$$\text{trace}(\alpha^{i+s}) = \text{trace}(\alpha^{j+s})$$

for all integers s with $0 \leq s \leq d$, then

$$\text{trace}(\alpha^s(\alpha^i - \alpha^j)) = 0, \quad 0 \leq s \leq d.$$

If $\alpha^i \neq \alpha^j$, this shows that the trace function would be identically 0 which is impossible: Otherwise, the polynomial $x + x^q + x^{q^2} + \ldots + x^{q^d}$ (which describes the trace function) has q^{d+1} zeroes, which is impossible. Therefore we must have $\alpha^i = \alpha^j$. This has the following interesting consequence: The periodic sequence $(a_i) = (\text{trace}(\alpha^i))$ with period $q^{d+1} - 1$ over $GF(q)$ contains every non-zero vector of $GF(q)^{d+1}$ as a "substring" $(a_i, a_{i+1}, \ldots, a_{i+d})$, $0 \leq i \leq q^{d+1} - 2$, exactly once: If we want to generate the sequence (a_i), it is basically enough to consider the recurrence relation corresponding to the minimum polynomial of the primitive element α. We generate a sequence (b_i) with an <u>arbitrary</u> vector to begin with. A suitable translate (b_{i+n}) of this sequence has to be the sequence $\text{trace}(\alpha^i)$, hence the indices i modulo $(q^{d+1} - 1)/(q-1)$ with $b_i = 0$ form a translate of the Singer difference set which consists of the elements whose trace is 0.

Example 3.1.3 It is fairly easy to construct even large Singer difference sets (easy modulo the knowledge of a primitive polynomial). We try to construct a Singer difference set with parameters $(40, 13, 4)$ corresponding to $PG(3, 3)$. A primitive polynomial of degree 4 over $GF(3)$ is, for instance, $x^4 + x + 2$. This polynomial describes a linear recurrence relation

$$a_{4+i} = 2a_{i+1} + a_i.$$

We choose the start vector $(0, 0, 0, 1)$ and obtain the following sequence:

$$0001002101112002201022110101212212012222220002\ldots$$

This gives rise to a difference set $D = \{0, 1, 2, 4, 5, 8, 13, 14, 17, 19, 24, 26, 34\}$ in $\mathbb{Z}_{40}$. The difference set consisting of the elements with trace 0 is $D + 1$.

Sometimes, sequences whith entries from a field K which satisfy a linear recurrence relation are called **linear shift register sequences**. A **linear shift register of degree** n over a field K is a "device" which is able to produce a sequence which satisfies a linear recurrence relation of degree n. More precisely, a **shift register** consists of n linearly ordered cells $D_0, \ldots, D_{n-1}$ which are able to store elements from K. The vector $(a_0, \ldots, a_{n-1})$ containing the elements stored in the D_i's is called a **state vector**. After certain time intervals, the element in the left-most cell D_0 becomes the output symbol, and the entries of every other cell move to the left. The new entry of D_{n-1} will be computed from the entries in the cells before they are shifted. We call the shift register **linear** if the new entry is $c_0 a_0 + c_1 a_1 + \ldots + c_{n-1} a_{n-1}$, where the c_i's are elements from K indendent of the state vector which is stored in the shift register. We can visualize the linear shift register used in Example 3.1.3 in a diagram as follows:

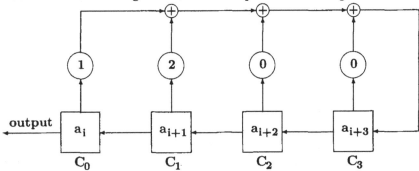

Linear shift registers are extremely easy to realize as a hardware device (at least in the binary case), and therefore linear shift register sequences are easy to generate, in particular, Singer difference sets are easy to generate. However, the situation becomes more complicated if the calculations have to be carried out in a finite field whose order is not a prime (see Jungnickel [101] for the problem of doing arithmetics in arbitrary finite fields). We refer the reader to Golomb [76] and Lüneburg [122] for a thorough investigation of shift registers. Moreover, the books of Jungnickel [101] as well as Lidl and Niederreiter [120] contain chapters about linear shift registers.

A (linear) shift register sequence over $GF(q)$ of period length $q^n - 1$ which can be generated on a shift register of degree n is called an m-**sequence** (m for maximal). The notion "maximal" is easy to explain: The period length of any sequence which is generated on a linear shift register of degree n is at most $q^n - 1$. We have seen that Singer difference sets are closely related to the m-sequences defined by the trace-function together with a primitive element α. It is not difficult to show that every m-sequence over $GF(q)$ with period $q^n - 1$ is equivalent to this m-sequence. (Here we call two periodic sequences (a_i) and (b_i) with period v **equivalent** if $a_i = b_{(si+t) \bmod v}$ for suitable integers s and t with s relatively prime to v. In particular, the sequences trace(α^i) and trace(β^i) are equivalent if both α and β are primitive elements in $GF(q^n)$.) Before we are going to prove this assertion, we will give another characterization of the linear complexity of a periodic sequence (see Section 1.2 for the original definition of the linear complexity).

Theorem 3.1.4 *Let (a_i) be a periodic sequence with period v over the field $GF(q)$. The linear complexity of (a_i) is the smallest degree of a linear shift register over $GF(q)$ which is capable to generate (a_i).*

Proof. We define

$$a^{(j)} := (a_j, a_{j+1}, \ldots, a_{j+v-1}),$$

(indices modulo v). If the linear complexity of the sequence is n, then there are n elements $c_0, \ldots, c_{n-1}$ such that

$$a^{(n)} = c_{n-1}a^{(n-1)} + \ldots + c_0 a^{(0)}.$$

This shows that the sequence (a_i) satisfies a linear recurrence relation of degree n. The same argument shows that any $n+1$ vectors $a^{(i)}, \ldots, a^{(i+n)}$ are linearly dependent if the sequence (a_i) satisfies a linear recurrence relation of degree n (superscripts modulo v). But then the rank of the matrix with rows $a^{(i)}$ is at most n, which finishes the proof. $\qquad\square$

Theorem 3.1.5 *Let (a_i) be an m-sequence over $GF(q)$ with period $q^n - 1$. If a_i satisfies a linear recurrence relation*

$$a_{n+i} = c_{n-1}a_{n+i-1} + c_{n-2}a_{n+i-2} + \ldots + c_0 a_i,$$

then the polynomial $m(x) := x^n - c_{n-1}x^{n-1} - \ldots - c_0$ is primitive. If α is a root of $m(x)$, then the sequence (a_i) is a translate of the sequence $(\text{trace}(\alpha^i))$. This shows that any two m-sequences over $GF(q)$ of period $q^n - 1$ are equivalent.

Proof. In contrast to the proof of Theorem 3.1.4, we define

$$a^{(j)} := (a_j, \ldots, a_{j+n-1}).$$

Let $\mathbf{C}$ be the $(n \times n)$-matrix

$$\mathbf{C} := \begin{pmatrix} 0 & 0 & 0 & \cdots & & c_0 \\ 1 & 0 & \cdots & \cdots & & c_1 \\ 0 & 1 & 0 & \cdots & & c_2 \\ \vdots & \ddots & \ddots & \ddots & & \vdots \\ 0 & \cdots & 0 & 1 & & c_{n-1} \end{pmatrix}.$$

The characteristic polynomial of $\mathbf{C}$ is $m(x)$. We have $a^{(i)}\mathbf{C} = a^{(i+1)}$ and $a^{(0)}\mathbf{C}^i = a^{(i)}$, hence the order of $\mathbf{C}$ is $q^n - 1$, and the minimum polynomial of $\mathbf{C}$ is $m(x)$, too (otherwise the subalgebra generated by $\mathbf{C}$ in the algebra of all $(n \times n)$-matrices has less than q^n elements). But this subalgebra is actually a field, and the linear algebra minimum polynomial of $\mathbf{C}$ is the same as the minimum polynomial of $\mathbf{C}$ over $GF(q)$ considered as an element in the field $GF(q^n)$.

Hence $m(x)$ is a primitive polynomial (since the order of $\mathbf{C}$ is $q^n - 1$). $\qquad\square$

The linear complexity of an m-sequence is just the degree of the corresponding recurrence relation. If $q = 2$, this is the same as the GF(2)-rank of an incidence matrix which describes the complement of the point-hyperplane design of PG($d, 2$).

Corollary 3.1.6 *The rank of an incidence matrix of the point-hyperplane design of PG($d, 2$) over GF(2) is $d+2$. The complementary design has GF(2)-rank $d+1$.*

Proof. The rank of the complementary design is $d + 1$ since the corresponding difference set "is" an m-sequence. If we replace the difference set by its complement, we replace a non-principal character $\chi(D)$ by $\chi(G - D) = -\chi(D)$ and $\chi_0(D) = (q^d - 1)/(q - 1) \neq 0$ by $\chi_0(G - D) = q^d = 0$ in GF(2). But the number of character values $\chi(D)$ different from 0 is the rank of the group invariant incidence matrix corresponding to D, see Proposition 1.2.16. $\qquad\square$

In Theorem 3.2.9, we will determine the ranks of the incidence matrices of the other point-hyperplane designs (equivalently, the dimensions of the ideals generated by the Singer difference sets), too. In connection with Corollary 3.1.6, the following question due to Dillon is of interest:

Problem 8 (Dillon [64]) Let D be a cyclic difference set in G with parameters (3.2) where q is a power of 2. Is it true that the ideal generated by D in GF(2)[G] (in other words, the code corresponding to D) <u>contains</u> an ideal which is generated by the complement of a classical Singer difference set?

One can rephrase the question in terms of shift registers: If a linear shift register generates a difference set with parameters (3.2), then the shift register would also be capable to generate a classical Singer difference set (for an appropriate initial start vector).

At this point, we will investigate some questions about Singer difference sets which are typically asked in connection with a series of difference sets. First of all: Are there difference sets with parameters (3.1) which describe designs which are different from the point-hyperplane designs of PG(d, q). The answer to this question is yes. We will describe such a construction (which is due to Gordon, Mills and Welch [77]) in Section 3.2.

Another question is whether the construction in Theorem 2.1.1 is unique, i.e., whether the only difference set which describes a point-hyperplane-design has to be equivalent to the difference set constructed above. The answer is yes if $q = 2$, but it seems to be unknown if $q > 2$.

Proposition 3.1.7 *Let D be a cyclic difference set with parameters $(2^{d+1} - 1, 2^d, 2^{d-1})$ corresponding to PG($d,2$). Then D is equivalent to the complement of a classical Singer difference set.*

Proof. The GF(2)-code of the difference set has dimension $d + 1$. Therefore we can generate the associated sequence on a shift register of size $d + 1$ and the sequence is an m-sequence. But these are unique (up to equivalence) according to Theorem 3.1.5. □

Moreover, one can ask questions about the structure of the group, i.e., are there other groups which contain difference sets with parameters (3.1). The answer is yes, there are some examples of non-abelian groups containing difference sets with parameters (3.1), see Gao and Wei [74] and Pott [146]. However, no abelian but non-cyclic example is known!

Now we replace the prime power q by an arbitrary non-negative integer s and ask whether difference sets with parameters

$$\left(\frac{s^{d+1} - 1}{s - 1}, \frac{s^d - 1}{s - 1}, \frac{s^{d-1} - 1}{s - 1}; s^{d-1} \right) \tag{3.4}$$

exist. In other words, we have a hypothetical two-parameter family of difference sets and ask for which parameters do difference sets exist. No example of such a difference set is known where s is not a prime power. But it seems that this question has not attrackted much attention which might be the reason for the deficit of non-existence and existence results. The case $d = 2$, however, has been investigated quite extensively: If $d = 2$, the parameters can be rewritten as

$$(s^2 + s + 1, s + 1, 1; s)$$

and the difference sets correspond to projective planes. It is one of the main topics of this monograph to study projective planes admitting regular or, more generally, "quasiregular" collineation groups using results from algebraic number theory.

3.2 The Gordon-Mills-Welch construction

Besides the classical Singer difference sets, only one <u>infinite</u> family of difference sets with the classical parameters is known which are not Singer difference sets.

Proposition 3.2.1 *Let R be an abelian (m, n, k, λ)-difference set in G relative to N, and let T be an abelian $(n/n', n', l, \mu)$-difference set in N relative to H. Then the set*

$$D := \{ (r_1 r_2 : r_1 \in R, r_2 \in T \}$$

is a difference set in G relative to H with parameters

$$(mn/n', n', kl, k\mu)$$

provided that

$$k\mu = l\lambda + \mu\lambda(n - n'). \tag{3.5}$$

Proof. Since D contains at most one element from each coset of N, the product RT considered in $\mathbf{Z}[G]$ has $0/1$-coefficients. Using the basic equations for R and T (see (1.7)), we obtain

$$
\begin{aligned}
(RT)(RT)^{-1} &= (k + \lambda(G - N))(l + \mu(N - H)) \\
&= lk + (l\lambda + \mu\lambda(n - n'))G + \\
&\quad + (k\mu - l\lambda + \mu\lambda(n' - n))N - k\mu H.
\end{aligned}
$$

Therefore RT describes a relative difference set with the desired parameters if and only if (3.5) holds.   □

Not many relative difference sets are known, and therefore not many relative difference sets are known which satisfy (3.5). We can take semiregular relative difference sets to get new semiregular examples in the same groups but where the orders of the forbidden subgroups become smaller. We will say more about semiregular relative difference sets in the next chapter.

Difference sets with the affine parameters provide us with another series of examples which satisfy (3.5): We take

$$
\begin{aligned}
m &= \frac{q^{st} - 1}{q^t - 1}, \\
n &= q^t - 1, \\
k &= q^{st-t}, \\
n' &= q - 1, \\
l &= q^{t-1}.
\end{aligned}
$$

Corollary 3.2.2 (Arasu, Pott, Spence [20]) *Let R be an abelian difference set in G relative to N with parameters*

$$
\left(\frac{q^{st} - 1}{q^t - 1}, q^t - 1, q^{st-t}, q^{st-2t} \right)
$$

(where q is not necessarily a prime power, although we do not know of any other example). Moreover, let T be an abelian difference set in N relative to H with parameters

$$
\left(\frac{q^t - 1}{q - 1}, q - 1, q^{t-1}, q^{t-2} \right).
$$

Then RT in $\mathbf{Z}[G]$ is a difference set in G relative to H with parameters

$$
\left(\frac{q^{st} - 1}{q - 1}, q - 1, q^{st-1}, q^{st-2} \right). \tag{3.6}
$$

□

Of course, these are the classical affine parameters, and the construction in Corollary 3.2.2 does not seem to be interesting. However, we can replace one of the difference sets R or T by an equivalent one, say, for instance, R' or T'. Then Corollary 3.2.2 shows that $R'T'$ is again a relative difference set. Interestingly enough, these examples are, in general, not equivalent to the original difference sets:

Theorem 3.2.3 *Let R be the classical cyclic affine difference set in G relative to N with parameters*

$$\left(\frac{q^{st}-1}{q^t-1}, q^t-1, q^{st-t}, q^{st-2t}\right)$$

and $s > 1$. Moreover, let T and T' be arbitrary relative difference sets in N relative to H with parameters

$$\left(\frac{q^t-1}{q-1}, q-1, q^{t-1}, q^{t-2}\right).$$

*If T and T' are not multiplier equivalent, then the relative difference set RT is not equivalent to RT'. These difference sets are called the **affine GMW-difference sets**.*

Proof. Assume that $RT' = (RT)^{(b)}x$ for some x in G and some integer b, i.e., suppose that RT and RT' are equivalent. We apply the projection epimorphism ρ from G onto G/N to obtain

$$|T|\ R = |T|\ R^{(b)}\rho(x)$$

or

$$R = R^{(b)}\rho(x) \quad \text{in } \mathbf{Z}[G/N].$$

Since the only multipliers of the classical affine difference sets are the powers of p (if $q = p^r$), see Corollary 3.1.2, the integer b must be a power of p modulo $(q^{st}-1)/(q-1)$, see Theorem 3.1.1, and therefore b is a multiplier of T, too (Corollary 1.3.6). Hence we obtain $RT' = RT^{(b)}y$ in $\mathbf{Z}[G]$. The group ring element R has an inverse in $\mathbf{Q}[G]$ (since we are in the regular case) and thus $T' = T^{(b)}y$. This contradicts our assumption that T' is not multiplier equivalent to T. □

This theorem shows that there are many non-equivalent relative difference sets with the classical parameters. It is a different (and apparently more difficult) question to decide whether the corresponding designs are isomorphic. Before we discuss this problem, we will describe the connection of our construction with the so called **GMW construction** of difference sets due to Gordon, Mills and Welch [77]. The group H in Corollary 3.2.2 has order $q-1$. We denote the canonical epimorphism $G \to G/H$ by μ. Note that $\mu(R)$ is a relative $((q^{st}-1)/(q-1), (q^t-1)/(q-1), q^{st-t}, q^{st-2t}(q-1))$-difference set, and $\mu(T)$ is a difference set with parameters $((q^t-1)/(q-1), q^{t-1}, q^{t-2}(q-1))$, and the product $\mu(R)\mu(T)$ is again a difference set with parameters $((q^{st}-1)/(q-1), q^{st-1}, q^{st-2}(q-1))$. This

difference set is, of course, a projection of the relative difference set in Corollary
3.2.2. This is precisely the construction of Gordon, Mills and Welch: Take
a relative difference set R_{GMW} and a difference set T_{GMW} (with appropriate
parameters), then the product $R_{GMW} \cdot T_{GMW}$ becomes a difference set. The
proof of Theorem 3.2.3 which shows that we get inequivalent difference sets if
we replace T_{GMW} by something which is not multiplier equivalent to T_{GMW}
remains true since the only multipliers of T_{GMW} are powers of p if T_{GMW} is
a projection of a classical affine difference set (provided that $t > 2$). Thus we
have:

Theorem 3.2.4 (Gordon, Mills, Welch [77]) *Let $d + 1 = st$, $s \neq 1$, $t > 2$.
Then the complement of a classical Singer difference set in G with parameters
(3.2) can be written as a product $R_{GMW} \cdot T_{GMW}$ in $\mathbf{Z}[G]$ (where the parameters
R_{GMW} and T_{GMW} are as above). Let r be an integer relatively prime to $(q^t -
1)/(q - 1)$ and which is not a power of the prime p ($q = p^a$). If T_{GMW} is
replaced by $T' := (T_{GMW})^r$, then $R_{GMW} \cdot T'$ describes a difference set with
parameters (3.2) which is not equivalent to the classical Singer difference set.
These difference sets are called the* **classical GMW-difference sets.** □

Corollary 3.2.5 *Let D_1 and D_2 be two GMW-difference sets constructed according to Theorem 3.2.4 using integers r_1 and r_2. If there is no integer i such
that $r_1 r_2^{-1} \equiv p^i \bmod (q^t - 1)/(q - 1)$, then D_1 and D_2 are not equivalent.* □

It is now quite a natural question to ask whether the designs corresponding
to inequivalent GMW-difference sets are isomorphic or not. This seems to be a
difficult question. We can give an answer for the case $p = 2$ using the GF(2)-rank
of the associated incidence matrices. Obviously, when the ranks are different,
the designs have to be non-isomorphic. In the first chapter of this monograph,
we have developed a method how to compute the rank of incidence matrices
of designs which have a description as a difference set. In order to apply this
method, we need another way to look at the GMW-construction. Let $d = st$ be
a composite number. It is well-known and also very easy to see that the trace
function from GF(q^d) onto GF(q) denoted by trace$_{d/1}$ can be decomposed if d
is composite

$$\text{trace}_{d/1} = \text{trace}_{t/1}(\text{trace}_{d/t})$$

where trace$_{d/t}$ denotes the trace function from GF(q^d) onto GF(q^t). As in the
GMW-construction, we assume that r is an integer relatively prime to $q^t - 1$.
Then we call the sequence (b_i) with period length $q^d - 1$ defined by

$$b_i := \text{trace}_{t/1}(\text{trace}_{d/t}(\alpha^i))^r \tag{3.7}$$

a **GMW-sequence** where α is a primitive element of GF(q^d). If $r = 1$, we just
get an m-sequence. The sequence (b_i) is periodic with period $q^d - 1$. Moreover,
it satisfies

$$b_{i+j(q^d-1)/(q-1)} = \gamma^j b_i \tag{3.8}$$

where γ is a primitive element in GF(q). The sequence "describes" a relative
difference set constructed according to Theorem 3.2.3.

Proposition 3.2.6 *Let (b_i) be a GMW-sequence (3.7), and let D be the affine GMW-difference set $RT^{r^{-1}}$. We assume that R is the affine difference set*

$$\{0 \le i \le q^d - 2 : trace_{d/t}(\alpha^i) \ne 0\}$$

and T is the affine difference set

$$\{0 \le i \le q^t - 2 : trace_{t/1}(\beta^i) = 1\}.$$

Here α denotes a primitive element of $GF(q^d)$, and $\beta := \alpha^{(q^d-1)/(q^t-1)}$ is a primitive element in $GF(q^t)$. Then

$$D = \{0 \le i \le q^d - 2 : b_i = 1\}.$$

Proof. We write the element $\alpha^i := \alpha^{i_1}\beta^{i_2}$ such that $trace_{d/t}(\alpha^{i_1}) = 1$ or 0 (note that it is always possible to write α^i in this form). We obtain

$$b_i = 1 \quad \Leftrightarrow \quad trace_{t/1}(\beta^{r i_2}) = 1.$$

This is the case if and only if i can be written as $i = i_1 + (q^d - 1)/(q^t - 1)i_2$ with $i_1 \in R$ and $i_2 \in T^{r^{-1}}$.   □

It is rather obvious that we can construct a GMW-difference set from a GMW-sequence if we just look at the first $(q^d - 1)/(q - 1)$ elements in the sequence (b_i), see (3.8):

$$\{0 \le i < \frac{q^d - 1}{q - 1} - 1 : b_i \ne 0\}$$

is a classical GMW-difference set.

We will now compute the linear complexities of the GMW-sequences. If $q = 2$, these complexities are the same as the ranks of the incidence matrices defined by the GMW-difference sets, and it turns out that these ranks are different for different choices of r. This will show that (at least in the case $q = 2$) inequivalent GMW-difference sets are also non-isomorphic. If $q \ne 2$, it is also possible to compute the complexities of the GMW-sequences, but now it is not so easy to relate the complexities of the sequences to the ranks of the designs: We take the sequence (b_i) with entries from $GF(q)$ and raise the entries to the $(q - 1)$-st power to obtain a $0/1$-sequence (c_i) with $c_i = b_i^{q-1}$. We know that we can think of the sequence (c_i) as an element in the group algebra $GF(q)[G]$ where $G = \langle g \rangle$ is a cyclic group of order $q^d - 1$:

$$C := \sum_{i=0}^{q^d-1} c_i g^i.$$

If we quotient the cyclic subgroup Q of order $q - 1$ (denote this projection by ψ) we obtain

$$\psi(C) = (q - 1) \sum_{i=0}^{(q^d-1)/(q-1)} c_i \psi(g^i)$$

since the coefficients of C are constant on cosets of Q (see (3.8)). The group algebra element $\sum_{i=0}^{(q^d-1)/(q-1)} c_i \psi(g^i)$ "is" a GMW-difference set. The number of characters χ from G into an extension field of GF(q) such that $\chi(C) \neq 0$ is the linear complexity of (c_i). There are two types of characters: Those, which are principal on Q and those which are not. If χ is non-principal on Q, then $\chi(C) = 0$ in view of Lemma 1.2.1. If χ is principal on Q, then χ is also a character of $\psi(G)$ and every character of $\psi(G)$ arises in this form. Hence the linear complexity is the number of characters χ of $\psi(G)$ such that $\chi(\psi(C)) \neq 0$. But this is precisely the rank of the difference set design that belongs to the GMW-difference set $\psi(C)$.

A similar problem arises if we want to compute the ranks of the divisible designs corresponding to the classical affine difference sets. These are closely related to m-sequences (a_i). The group algebra element corresponding to the difference set is now

$$\sum_{i=0}^{q^d-2} c_i g^i$$

where $c_i = 1$ if $a_i = 1$ and $c_i = 0$ otherwise. Hence the rank of the incidence matrix corresponding to a classical affine difference set is the linear complexity of the sequence $1 - (a_i - 1)^{q-1}$. We can summarize this in the following proposition:

Proposition 3.2.7 *Let q be a power of the prime p. Suppose $d = st$, and let (a_i) denote an m-sequence over GF(q) with period $q^d - 1$, and let (b_i) denote a GMW-sequence (3.7). The following table translates the linear complexities of some sequences derived from (a_i) and (b_i) into the ranks of certain designs corresponding to cyclic GMW-difference sets:*

sequence	difference set
(a_i^{q-1})	Singer difference set
(b_i^{q-1})	GMW-difference set
$(1 - (a_i - 1)^{q-1})$	classical affine difference set
$(1 - (b_i - 1)^{q-1})$	affine GMW-difference set

□

We will compute the linear complexities of some of the sequences in this table. The GMW-sequence (b_i) can be defined by a polynomial $g \in$ GF(q)$[x]$ such that

$$b_i = g(\alpha^i)$$

where α is a primitive element of GF(q^d). Using the trace description of b_i, this polynomial is

$$g := f^r + (f^r)^q + (f^r)^{q^2} + \ldots + (f^r)^{q^{t-1}} \tag{3.9}$$

where

$$f := x + x^{q^t} + x^{q^{2t}} + x^{q^{(s-1)t}}.$$

Let us define one more polynomial

$$h := x + x^{q^t} + x^{q^{2t}} + x^{q^{(s-1)t}} - 1.$$

We define the **support** of a polynomial $f = \sum f_i x^i$ to be the number of non-zero coefficients f_i. Corollary 1.2.17 shows that the sizes of the supports of the polynomials f^{q-1}, $1 - h^{q-1}$, g^{q-1} and $1 - (g-1)^{q-1}$ modulo $x^{q^d-1} - 1$ are the linear complexities of the sequences we are interested in.

We begin to consider the polynomials f^r and h^r. To simplify the notation a bit, we assume $t = 1$ (hence $d = s$) without loss of generality. Let us write $q = p^e$ and assume $r \leq q - 1$. Again, this is not a restriction since the range of $x + x^q + \ldots + x^{q^{s-1}}$ (considered as a function on $GF(q^d)$) is $GF(q)$ and $y^{q-1} = 1$ for $y \in GF(q)$. It is important for the following arguments that the degrees of f^r and h^r are at most $q^d - 2$, which shows that the polynomials f^r and h^r reduced modulo $x^{q^d-1} - 1$ are the same as f^r and h^r. The only exceptions are f^{q-1} and h^{q-1} with $d = 1$. We will exclude these (rather uninteresting) cases from now on: The linear complexities are 1 in the first case and $q - 2$ or $q - 1$ in the second case, depending on whether q is even or odd ($h^{q-1} = (x-1)^{q-1} \equiv x + \ldots + x^{q-2} \mod (x^{q-1} - 1)$ if q is even).

We write

$$r = \sum_{i=0}^{e-1} r_i p^i \tag{3.10}$$

where the r_i's are integers in the interval $[0, p-1]$. Then the r_i's are uniquely determined (p-ary expansion of r). We write

$$f^r = \prod_{i=0}^{e-1} (x^{p^i} + x^{p^{e+i}} + \ldots + x^{p^{(d-1)e+i}})^{r_i} =: \sum b_j x^j. \tag{3.11}$$

First, we consider

$$(x^{p^i} + x^{p^{e+i}} + \ldots + x^{p^{(d-1)e+i}})^{r_i} =: \sum_k c_k^{(i)} x^k.$$

The coefficient $c_k^{(i)}$ can only be different from 0 if

$$k = \sum_{m=0}^{d-1} p^{i+me} s_m \quad \text{with} \quad \sum_m s_m = r_i, \quad 0 \leq s_m < p. \tag{3.12}$$

The $c_k^{(i)}$'s are multinomial coefficients which are definitely not zero since $r_i < p$, therefore, we have $c_k^{(i)} \neq 0$ if and only if (3.12) holds. We define $T_i := \{k : c_k^{(i)} \neq 0\}$, and note that T_i consists of integers whose p-ary expansion contains only exponents congruent i modulo e. Since the p-ary representation is unique, we obtain

$$\sum_{m=0}^{d-1} p^{i+me} s_m = \sum_{m=0}^{d-1} p^{i+me} t_m$$

if and only if $s_m = t_m$ for $m = 0, \ldots, d-1$. Thus the order of T_i is the number of possibilities to write r_i as a sum of d non-negative integers. This number (denoted by $n(d, r_i)$) is

$$n(d, r_i) = \binom{d + r_i - 1}{d - 1},$$

see any textbook on (elementary) Combinatorics, for instance Brualdi [37]. Now we get $b_j \neq 0$ if and only if we can write j as a sum

$$j = \sum_{i=0}^{e-1} t_i \quad \text{with} \quad t_i \in T_i.$$

Since the p-ary representation is unique, we obtain again $\sum t_i = \sum s_i$ if and only if $t_i = s_i$. Therefore, the number of non-zero coefficients b_j in (3.11) is

$$\prod_{i=0}^{e-1} n(d, r_i).$$

Similar arguments work if we look at h^r. In this case, we get x^j with a non-zero coefficient in

$$(x + x^q + \ldots + x^{q^{d-1}} - 1)^r$$

if and only if

$$j = \sum_{n=0}^{de-1} s_n p^n \quad \text{with} \quad \sum_{m \in T_i} s_m \leq r_i, 0 \leq s_m < p.$$

The formula for the number of possibilities to write the numbers less than or equal to r_i as a sum of d non-negative integers is $n(d+1, r_i)$. We can summarize this in the following Theorem.

Theorem 3.2.8 (Antweiler, Bömer [1]) *Let $q = p^e$ be a power of the prime p, and let $r = \sum_{i=0}^{e-1} r_i p^i$, $r_i \in [0, p)$. The trace denotes the usual trace function from $GF(q^d)$ onto $GF(q)$. Moreover, let α be a primitive element in $GF(q^d)$. Then the following holds:*

(a) *The linear complexity of the sequence $(trace(\alpha^i)^r)$ is*

$$\prod_{i=0}^{e-1} \binom{d + r_i - 1}{d - 1}.$$

(b) *The linear complexity of the sequence $((trace(\alpha^i) - 1)^r)$ is*

$$\prod_{i=0}^{e-1} \binom{d + r_i}{d}$$

(except $d = 1$, $r = q - 1$ and q is even, where the complexity is $q - 2$, or $d = 1$, $r = q - 1$ and q is odd, where the complexity is $q - 1$).

In general, the sequences which occur in this theorem do not have 0/1-coefficients and so they are not too interesting in connection with difference set problems. On the other hand, it is of some interest to determine the linear complexities of "interesting" sequences, which we have done in this theorem. Part (1) of the theorem is contained in [1]. I do not know a reference for the second part. I have stated it only because it yields a dimension formula for the classical affine designs:

Corollary 3.2.9 (MacWilliams, Mann [130]) *Let $q = p^e$ be a power of the prime p and let d be an integer $(d \geq 2)$.*

(a) *The $GF(p)$-rank of the point-hyperplane design of $PG(d-1, q)$ $(d \geq 3)$ is*

$$\binom{d+p-2}{d-1}^e + 1.$$

(b) *The $GF(p)$-rank of the point-hyperplane design of $AG(d, q)$ $(d \geq 2)$ is*

$$\binom{d+p-1}{d}^e.$$

Proof. (a) The $GF(p)$-rank of the complement of the point-hyperplane design is the same as the $GF(p)$-linear complexity of the sequence defined via the polynomial f^{q-1}. In order to get the rank of the point-hyperplane design, we must replace f^{q-1} by $1 - f^{q-1}$ (we go to the complementary design). The support of this polynomial contains exactly one more element than the support of f^{q-1} (the coefficient of x^0) which gives the rank formula.

(b) The polynomial that describes the cyclic affine difference set (or the corresponding sequence) is $1 - h^{q-1}$. Note that the coefficient of x^0 in h^{q-1} is 1, hence the dimension of the affine difference set design is $\binom{d+p-1}{d}^e - 1$. We extend the group invariant matrix corresponding to the affine difference set first by a column that consists entirely of 0's and then by the all-one row vector. The row space V generated by this matrix contains the characteristic vectors of all the hyperplanes of $AG(d, q)$, and, vice versa, V is contained in the space generated by the hyperplanes. Its rank is exactly one more than the rank of the group invariant matrix of the affine difference set. This proves the second formula. □

We are now going to compute the complexities of the GMW-difference sets. We have to determine the size of the support of the polynomial (3.9). Of course, we do not assume $t = 1$ here! Again, we write the p-ary expansion of r as

$$r = \sum_{i=0}^{et-1} r_i p^i.$$

We assert that the supports of $f^r, (f^r)^q, \ldots$ are disjoint. Assume otherwise that the coefficient of x^j in $(f^r)^{q^a}$ and in $(f^r)^{q^b}$ is non-zero $(a \neq b, a, b < t)$. We write

$$j \equiv \sum_{i=0}^{et-1} \sum_{m=0}^{s-1} a_{etm+i} p^{etm+i+ea} \equiv \sum_{i=0}^{et-1} \sum_{m=0}^{s-1} b_{etm+i} p^{etm+i+eb} \bmod (p^{est}-1) \quad (3.13)$$

with

$$\sum_{m=0}^{s-1} a_{etm+i} = \sum_{m=0}^{s-1} b_{etm+i} = r_i. \quad (3.14)$$

Consider equation (3.13) modulo $p^{et} - 1$. Using (3.14), we get

$$\sum_{i=0}^{et-1} r_i p^{i+ea} \equiv \sum_{i=0}^{et-1} r_i p^{i+eb} \bmod (p^{et} - 1),$$

hence

$$r p^{ea} \equiv r p^{eb} \bmod (p^{et} - 1),$$

i.e. $s = t$ (since $a, b < t$, and r is relatively prime to $p^{et} - 1$). This proves the following theorem:

Theorem 3.2.10 (Antweiler, Bömer [1]) *Let $d = st$ and let $q = p^e$ be a power of the prime p. Moreover, let $r = \sum_{i=0}^{et-1} r_i p^i$, $r_i \in [0, p)$, be an integer relatively prime to $q^t - 1$. Then the linear complexity of the sequence (b_i) defined by*

$$b_i := trace_{t/1}(trace_{d/t}(\alpha^i))^r$$

(where α is a primitive element of $GF(q^d)$) is

$$t \cdot \prod_{i=0}^{et-1} \binom{d + r_i - 1}{d - 1}.$$

$\square$

We note that this theorem is a slight generalization of what Antweiler and Bömer have stated. They have assumed $q = p$. We are interested in Theorem 3.2.10 since it immediately gives the dimensions of certain GMW-difference set codes, namely in the case $q = 2$:

Corollary 3.2.11 *The linear complexity of a GMW-sequence with $q = 2$ is*

$$t \cdot d^{\mathrm{wght}(r)}$$

where $wght(r)$ denotes the number of non-zero digits in the 2-ary expansion of r. $\square$

If we want to find dimension formulae for GMW-difference sets with $q \neq 2$, we have to find the support size of the polynomial g^{q-1} considered modulo $x^{q^d-1} - 1$. In principal, this is always possible, however, I do not know a <u>nice</u> formula.

Antweiler and Bömer [1] have considered sequences defined over $GF(p)$ via the polynomial g^s where s is relatively prime to $p-1$. They were not interested in the 0/1-sequences that can be obtained from the GMW-sequences (in which case we have to choose $s = p - 1$).

It would be also interesting to find dimension or rank formulae for the affine GMW-difference sets constructed in Theorem 3.2.3 as well. Again, this is possible if $q = 2$. But in this case the affine difference sets are actually difference sets in the usual sense, and their dimensions are determined by Corollary 3.2.11.

The following table gives some computational results about the ranks of the designs corresponding to GMW-difference sets. Let (b_i) be the *GMW*-sequence (3.7). We denote the linear complexity of (b_i) by L_1, the linear complexity of (b_i^{q-1}) by L_2 and the complexity of $(1 - (1 - b_i)^{q-1})$ by L_3. Note that L_2 and L_3 are the ranks of the designs corresponding to GMW-difference sets and affine GMW-difference sets. We include the values for $q = 2$ and the values L_1, although we have already obtained easy formulae for these cases.

q	d	t	r	L_1	L_2	L_3
2	6	2	1	6	6	6
2	6	3	1	6	6	6
			3	12	12	12
2	8	2	1	8	8	8
2	8	4	1	8	8	8
			7	32	32	32
2	10	2	1	10	10	10
2	10	5	1	10	10	10
			3	20	20	20
			5	20	20	20
			7	40	40	40
			11	40	40	40
			15	80	80	80
2	12	2	1	12	12	12
2	12	3	1	12	12	12
			3	48	48	48
2	12	4	1	12	12	12
			7	108	108	108
2	12	6	1	12	12	12
			5	24	24	24
			11	48	48	48

q	d	t	r	L_1	L_2	L_3
			13	48	48	48
			23	96	96	96
			31	192	192	192
3	4	2	1	4	10	14
			5	12	10	22
3	6	2	1	6	21	27
			5	36	21	57
3	6	3	1	6	21	27
			5	18	39	57
			7	18	45	63
			17	54	63	117
3	8	2	1	8	36	44
			5	80	36	116
3	8	4	1	8	36	44
			7	24	136	169
			11	24	200	224
			13	32	324	356
			17	72	200	272
			23	72	136	208
			41	96	36	132
			53	216	324	540
4	4	2	1	4	16	24
			7	16	16	48
4	6	2	1	6	36	48
			7	54	36	144
4	6	3	1	6	36	49
			5	12	144	169
			11	24	36	85
			13	24	144	193
			23	48	36	133
			31	96	144	337
5	4	2	1	4	35	69
			7	12	35	117
			13	24	35	117
			19	40	35	165
5	6	2	1	6	126	209
			7	36	126	538
			13	120	126	587

q	d	t	r	L_1	L_2	L_3
			19	300	126	916
5	6	3	1	6	126	209
			3	12	378	622
			7	18	330	745
			9	30	330	887
			11	18	390	823
			13	36	378	728
			17	36	390	965
			19	60	294	835
			21	30	426	803
			23	60	390	859
			33	48	294	1097
			37	54	594	1247
			39	90	330	817
			43	96	426	739
			47	90	342	910
			49	150	342	1016
			63	108	126	715
			69	180	390	857
			73	180	390	857
			99	300	594	1933
7	4	2	1	4	84	209
			5	12	84	379
			11	20	84	379
			13	28	84	449
			17	24	84	417
			19	36	84	401
			41	84	84	609
8	4	2	1	4	64	124
			5	8	64	232
			11	16	64	328
			13	16	64	220
			23	32	64	232
			31	64	64	448
9	4	2	1	4	100	224
			7	12	100	452
			11	12	100	588
			13	16	100	560
			17	36	100	500
			23	36	100	460

q	d	t	r	L_1	L_2	L_3
			41	48	100	544
			53	108	100	880

Let us now return for a moment to the decomposition of a Singer difference set as a product of a difference set and a relative difference set. Do we know more examples of (relative) difference sets that can be used in order to find new difference sets with the classical Singer parameters? Parametrically, the only known difference sets which can be used are the Paley-Hadamard difference sets with parameters $(q-1, (q-1)/2, (q-3)/4)$ whenever q is a Mersenne prime $(q = 2^m - 1)$. Using the twin prime construction, it might also work if the product of two twin primes is of the form $2^m - 1$. No other possibilities are known to me. Therefore it might be a good idea to look at the relative difference set part of the decomposition. The idea is to replace the relative difference set T by a "completely" different difference set and not just by an inequivalent one. We know that the only underlineknown examples of regular relative difference sets have the parameters of the classical affine sets. The only known constructions which yield infinitely many examples are the classical one and the construction in Theorem 3.2.3. But using the affine GMW-difference sets in the original GMW-construction is not too interesting since we can get the same examples using just the original GMW-construction. Therefore, we have to look for another infinite family of relative difference sets or, in a first step, we should try to find at least a few examples. The first interesting case is the case of relative $(73, 7, 64, 8)$-difference sets extending the complement of a classical Singer difference set with parameters $(73, 64, 56)$. With such a difference set it would be possible to construct (new) $(511, 255, 127)$-difference sets. Unfortunately, a computer search has shown that there is only one (abelian) relative $(73, 7, 64, 8)$-difference set (up to equivalence).

What can we say about this search? The underlying difference set is the complement of a classical Singer difference set, and it has parameters $(73, 64, 56)$. Using the multiplier 2, it is easy to check that this difference set is unique (up to equivalence). The cyclic group of order 73 has 8 orbits of size 9 under the map $x \mapsto 2x$ and one singleton orbit $\{0\}$. Any of the size 9 orbits is a difference set, hence we can assume that the relative difference set "lifts" a complement of $\{1, 2, 4, 8, 16, 32, 37, 55, 64\}$ in $\mathbf{Z}_{73}$. Using Corollary 1.3.6 we know that the multiplier 2 lifts to a multiplier of the relative difference set, say R. We may assume that R is fixed by the multiplier 2. Let us choose generators for the size 9 orbits in the difference set D, say $g_0, g_1, \ldots, g_7$. Each of these elements has to be combined with exactly one element x in $\mathbf{Z}_7$. The element 0 in $\mathbf{Z}_{73}$ has to be lifted to $(0, 0)$ in $\mathbf{Z}_{73} \times \mathbf{Z}_7$, otherwise $(0, x)$ and $(0, 2x)$ are both elements in R yielding the difference $x - 2x$ in $\mathbf{Z}_7$. Let us define

$$a \; := \; \text{number of orbits in } \mathbf{Z}_{73} \text{ combined with } 0,$$

$$b \; := \; \text{number of orbits in } \mathbf{Z}_{73} \text{ combined with } 1, 2 \text{ or } 4,$$

$$c \; := \; \text{number of orbits in } \mathbf{Z}_{73} \text{ combined with } 3, 5 \text{ or } 6.$$

If we apply a character χ principal on $\mathbf{Z}_{73}$ but non-principal on $\mathbf{Z}_7$, we obtain

$$
\begin{aligned}
\chi(R)\overline{\chi(R)} \;=\; & (1 + 9a + 3b(\zeta + \zeta^2 + \zeta^4) + 3c(\zeta^3 + \zeta^5 + \zeta^6)) \\
& \cdot(1 + 9a + 3b(\zeta^3 + \zeta^5 + \zeta^6) + 3c(\zeta + \zeta^2 + \zeta^4)) = 64
\end{aligned}
$$

(where ζ is a primitive 7-th root of unity). Using this it is easy to check that $a = 1$, $b = 4$, $c = 2$ or $a = 1$, $b = 2$, $c = 4$. We choose the first of these solutions, the second one is equivalent. We have seen that $\chi(R)\overline{\chi(R)} = 64$ holds for one character χ which is principal on $\mathbf{Z}_{73}$ and non-principal on $\mathbf{Z}_7$ if we lift our difference set according to one of the solutions above. But then the same is true for all algebraic conjugates of χ, i.e., for characters $\chi' : G \to \mathbf{Q}(\zeta)$ of the form $\chi'(g) = \sigma(\chi(g))$ for some Galois automorphism σ of $\mathbf{Q}(\zeta)$ since σ commutes with complex conjugation and fixes 64. There are 6 characters conjugate to χ.

Now we try to find x_i's in $\mathbf{Z}_7$ and combine them with the g_i's to get a subset R of $\mathbf{Z}_{73} \times \mathbf{Z}_7$, invariant under the map $x \to 2x$, such that $\chi(R)\overline{\chi(R)} = 64$ for a character χ of order $73 \cdot 7$. Then we have found our difference set: We know that $\chi(R)\overline{\chi(R)} = 64$ holds for the $72 \cdot 6$ algebraic conjugates of the character of order $73 \cdot 7$. We know that the same holds for the 6 characters principal on $\mathbf{Z}_{73}$ but non-principal on $\mathbf{Z}_7$. Finally, we know $\chi(R)\overline{\chi(R)} = 8$ for all non-principal characters which are principal on $\mathbf{Z}_{73}$ since the projection of the set R onto $\mathbf{Z}_{73}$ is a difference set of order 8. The value $\chi_0(R)$ is 64 by construction. So the character values are what they are supposed to be, therefore R is a relative difference set with the desired parameters. To find R, we may assume without loss of generality that at least one of the elements $g_1, \ldots, g_7$ gets combined with $1 \in \mathbf{Z}_7$: If $(g_i, x) \in R$, then (g_i, tx) is an equivalent difference set (provided that $t \neq 0$). It is too much to fix the orbit that gets combined with 0, too, since we have already assumed that our difference set is the complement of $\{1, 2, 4, 8, 16, 32, 37, 55, 64\}$ and hence we cannot apply automorphisms of $\mathbf{Z}_{73}$, again. We see that we have to check only

$$
7 \cdot 6 \cdot \binom{5}{3} \cdot 3^3 \cdot 3^2
$$

possibilities and each possibility requires checking the absolute value of just one character. This is much easier than trying to combine orbits and then checking the list of differences. As mentioned earlier, the result of this search has shown that there is, up to equivalence, only one relative difference set with the parameters $(73, 7, 64, 8)$.

A (similar) computer search for relative $(85, 3, 64, 16)$-difference sets lifting the classsical $(85, 64, 48)$-Singer difference set gave exactly two solutions. We can summarize our results in the following proposition:

Proposition 3.2.12 *The following table contains all possible cyclic liftings of the classical Singer difference sets with $k \leq 50$ or $k \leq 64$ and n odd.*

Cyclic relative difference sets with classical parameters

(7,2,4,1)

2	3	5	13

(13A, 2, 9, 3)

4	5	7	10	11	12	15	19	21

(13B, 2, 9, 3)

5	8	15	17	19	20	23	24	25

(21, 6, 16, 2)

0	2	4	5	8	16	17	32	54	64	73	83
90	97	103	108								

(31A,2,16,4)

0	5	9	11	13	14	21	28	38	41	49	50
51	53	56	57								

(31B,2,16,4)

0	7	10	11	13	18	19	20	21	25	36	40
45	53	57	59								

(31A, 4, 25, 5)

0	1	5	13	23	25	39	42	58	64	65	68
69	71	72	77	79	86	88	92	97	107	112	113
115											

(31B, 4, 25, 5)

0	1	5	9	25	34	43	45	46	48	64	69
72	83	84	91	97	99	101	106	112	113	116	119
123											

(40A, 2, 27, 9)

0	4	7	8	10	12	13	17	21	24	26	28
29	30	31	36	37	39	51	56	59	62	63	72
73	74	78									

(40B, 2, 27, 9)

0	4	7	10	11	12	16	19	21	22	28	29
30	32	33	34	36	38	48	53	57	63	64	66
71	77	79									

(40C, 2, 27, 9)

0	13	16	17	23	26	31	32	37	39	44	47
48	50	51	52	59	61	62	64	68	69	70	73
74	76	78									

```
(57A, 6, 49, 7)
  0    10    18    22    26    45    52    70    86    88    90   110
126   129   137   139   148   153   154   165   173   182   185   187
188   195   198   208   215   219   236   248   260   269   271   274
275   278   283   284   288   289   290   306   313   315   320   321
339

(57B, 6, 49, 7)
  0     8    11    23    27    33    34    36    37    39    43    50
 54    55    56    72    74    77    86   101   103   108   110   116
118   128   136   142   161   162   166   176   189   197   201   206
212   231   238   249   252   259   260   268   273   297   298   301
310

(73,7,64,8)
  0    27    43    47    54    86    87    94   107   108   119   172
174   177   179   185   188   195   197   205   214   216   229   231
238   241   255   269   277   279   299   309   315   344   345   348
353   354   358   370   371   376   383   390   394   395   405   410
413   428   432   441   447   453   458   462   476   479   482   495
503   507   509   510

(85A, 3, 64,16)
  1     2     4     8    13    16    23    25    26    32    35    37
 39    41    45    46    47    50    52    57    64    67    70    73
 74    75    78    82    90    92    94   100   104   105   113   114
121   128   134   139   140   145   146   147   148   150   151   156
161   164   165   180   184   188   197   200   201   203   208   210
226   228   229   242

(85B, 3, 64,16)
  1     2     4     5     8    10    13    16    20    25    26    27
 29    32    35    40    47    50    52    54    58    63    64    65
 67    70    71    80    94    99   100   104   108   116   121   126
128   130   134   140   141   142   145   151   159   160   161   163
177   188   198   200   203   207   208   209   216   229   231   232
242   243   249   252
```

Proof. If $k \leq 50$, this is a computer result of Lam, see [114]. In the range between 50 and 64, there are only four interesting cases with parameters

$$(85, 64, 48), \quad (127, 63, 31), \quad (127, 64, 32), \quad (73, 64, 56).$$

I have mentioned my computer search with $n = 3$ in the first and $n = 7$ in the last case. The only thing which we have to exclude is the existence of a (cyclic) relative $(127, 31, 63, 1)$-difference set. But this case can be killed using Theorem 2.4.6 with $u = 1$ and $p = 3$: It turns out that $3^{315} \equiv -1 \bmod 127 \cdot 31$. □

Recall that one motivation to look for relative difference sets came from the decomposition of the Singer difference sets into a relative and a usual difference

set. Unfortunately, we cannot get a new infinite series of difference sets with
the classical Singer parameters which is inequivalent to the classical examples.
The following table gives our present knowledge about "sporadic" difference sets
with the classical parameters $(2^{d+1} - 1, 2^d - 1, 2^{d-1} - 1; 2^{d-1})$ and $d \le 8$. The
list is complete in the sense that there are no other inequivalent examples in
this range. The cases $d > 8$ are still open. If we replace 2 by another prime,
not much seems to be known: The smallest parameter set where inequivalent
examples occur is $(121, 40, 13; 27)$, see the table in Baumert's book [26]. This
book can also serve as a reference for the following table if $d \le 6$. The case $d = 7$
is in Cheng [45], the case $d = 8$ in Dreier and Smith [65] and Bacher [27]. We
have listed just representatives of the orbits under the multiplier $x \mapsto 2x$, i.e.,
the first difference set is, for instance, $\{1, 2 \cdot 1, 2^2 \cdot 1\}$. A "?" indicates that the
difference set is not yet a member of an infinite family.

Cyclic $(2^{d+1} - 1, 2^d - 1, 2^{d-1} - 1; 2^{d-1})$-difference set, $d \le 8$	
2	1	Singer
3	0 1 5	Singer
4	1 3 15	Singer
	1 5 7	Paley
5	0 1 3 7 9 13 27	Singer
	0 1 3 5 9 23 27	GMW
6	1 3 7 9 15 27 29 31 47	Singer
	1 9 11 13 15 19 21 31 47	Paley
	1 3 5 7 19 23 27 47 63	Hall
	1 3 7 9 13 19 27 31 47	?
	1 3 5 9 15 19 27 29 55	?
	1 3 5 11 19 21 27 29 63	?
7	0 1 3 7 13 17 19 23 25 27 31 37 45 51 59 85 111 119 127	Singer
	0 7 11 13 17 19 23 27 39 43 47 51 53 55 63 85 95 111 119	GMW
	0 1 3 7 13 15 17 25 29 37 39 43 47 51 53 61 63 85 119	?
	0 7 9 11 15 17 19 25 29 31 37 43 51 53 61 63 85 87 119	?
8	1 7 13 17 21 23 31 35 37 39 51 53 55 59 61 75 77 79 83 85 91 95 103 109 123 183 187 219 223	Singer
	1 5 23 27 29 31 37 39 41 47 55 59 63 75 77 79 85 87 95 109 117 123 171 175 183 191 219 223 255	GMW
	1 3 5 7 13 15 19 21 23 25 27 37 39 45 55 57 59 63 77 83 91 107 117 123 183 191 219 239 255	?
	1 3 5 7 9 11 17 29 31 43 53 55 57 61 63 75 77 91 95 109 117 123 125 171 175 183 219 239 255	?
	1 3 5 7 9 11 17 29 31 43 53 55 57 61 63 75 77 91 95 109 117 123 125 171 175 183 219 239 255	?

We will finish our discussion of GMW-sequences and GMW-difference sets
with the computation of the linear complexity of a few more sequences related
to GMW-sequences. We want to replace the non-zero entries b_i in GF$(q)$by
elements $\varphi(b_i)$ from a field K which can be different from GF(q). We begin with
an apparently different problem. Suppose that $\mathbf{M}$ is an incidence matrix of a
symmetric divisisble $(m, n, k, 0, \lambda)$-design. We label the rows and columns of $\mathbf{M}$

in such a way that the n points (resp. blocks) corresponding to the rows (resp. columns) $in, in + 1, \ldots, in + n - 1$ form a point (resp. block) class. Then $\mathbf{M}$ can be considered as an $m \times m$-matrix consisting of $n \times n$-matrices. Each of these small matrices is a permutation matrix. If we multiply each of it by an $n \times n$-matrix $\mathbf{A}$, we get a matrix

$$\mathbf{M}' = \mathbf{M} \cdot (\mathbf{I}_m \otimes \mathbf{A}).$$

If the rank of $\mathbf{M}$ is mn, then the rank of $\mathbf{M}'$ is obviously $m \cdot \text{rank}(\mathbf{A})$.

What is the connection with GMW-sequences (b_i) over $GF(q)$ with period $q^d - 1$? We know that the linear complexity of (b_i) is the rank of the circulant matrix $\mathbf{C}$ with first row $(b_i)_{i=0,\ldots,q^d-2}$. Let us change this circulant matrix a little bit. We rearrange the rows and columns of $\mathbf{C}$ according to their residue modulo $(q^d - 1)/(q - 1)$. The row (and column) number $i + j(q^d - 1)/(q - 1)$ with $0 \leq j < q - 1$ becomes the row (and column) number $j + i(q - 1)$. Thus the matrix $\mathbf{C}$ becomes a matrix $(c_{i,j})$ consisting of m^2 small circulant matrices $(m = (q^d - 1)/(q - 1))$ of size $q - 1$. Now we just look at the entries 1 in this matrix and replace every other entry by 0. Denote the result by $\mathbf{D}$, i.e., $\mathbf{D}$ is an incidence matrix of the divisible design associated with the GMW-sequence, see Theorem 3.2.3. Assume that $b_{i+j(q^d-1)/(q-1)} = \gamma^x$ where γ is a primitive element in $GF(q)$. We get $c_{0,i(q-1)+j} = \gamma^x$. Let us look at the small circulant matrix $\mathbf{C}^{(i)}$ with first row $(c_{0,i(q-1)+j})_{j=0,\ldots,q-2}$, and let $\mathbf{D}^{(i)}$ be the corresponding matrix where we have replaced the entries $\neq 1$ by 0. Now we replace each entry γ^x in $\mathbf{C}^{(i)}$ by $\varphi(\gamma^x)$ and replace 0 by $\varphi(0) := 0$. It is easy to check that this matrix is

$$\mathbf{D}^{(i)}\mathbf{A}$$

where $\mathbf{A}$ is the circulant matrix with first <u>column</u> $(\varphi(\gamma^j))_{j=0,\ldots,q-2}$: This is true since the matrix $\mathbf{C}^{(i)}$ is not only circulant but also satisfies

$$c_{0,i(q-1)+j+1} = \gamma c_{0,i(q-1)+j}$$

because of (3.8). Thus we have proved:

Theorem 3.2.13 *Let (b_i) be a GMW-sequence with entries in $GF(q)$ and period $q^d - 1$. Suppose that K is a field whose characteristic does not divide q, and let φ be a mapping $GF(q) \to K$ with $\varphi(0) = 0$. Then the linear complexity of the sequence $(\varphi(b_i))$ (over K) is $(q^d - 1)/(q - 1) \cdot l$ where l is the K-rank of the circulant matrix with first row $(\varphi(1), \varphi(\gamma), \varphi(\gamma^2), \ldots, \varphi(\gamma^{q-1}))$ (γ is a generator of $GF(q)$).* $\square$

Example 3.2.14 We take $q = 4$, $d = 2$ and let $GF(4) = \{0, 1, \alpha, 1 + \alpha\}$ with primitive element α. Then the sequence

$$(0, 1, 1, \alpha, 1, 0, 1 + \alpha, 1 + \alpha, 1, 1 + \alpha, 0, \alpha, \alpha, 1 + \alpha, \alpha, \ldots)$$

is an m-sequence (and hence a GMW-sequence). The matrix $\mathbf{D}$ is

$$\mathbf{D} = \left(\begin{array}{ccc|ccc|ccc|ccc|ccc}
0 & 0 & 0 & 1 & 0 & 0 & 1 & 0 & 0 & 0 & 1 & 0 & 1 & 0 & 0 \\
0 & 0 & 0 & 0 & 1 & 0 & 0 & 1 & 0 & 0 & 0 & 1 & 0 & 1 & 0 \\
0 & 0 & 0 & 0 & 0 & 1 & 0 & 0 & 1 & 1 & 0 & 0 & 0 & 0 & 1 \\ \hline
0 & 1 & 0 & 0 & 0 & 0 & 1 & 0 & 0 & 1 & 0 & 0 & 0 & 1 & 0 \\
0 & 0 & 1 & 0 & 0 & 0 & 0 & 1 & 0 & 0 & 1 & 0 & 0 & 0 & 1 \\
1 & 0 & 0 & 0 & 0 & 0 & 0 & 0 & 1 & 0 & 0 & 1 & 1 & 0 & 0 \\ \hline
0 & 0 & 1 & 0 & 1 & 0 & 0 & 0 & 0 & 1 & 0 & 0 & 1 & 0 & 0 \\
1 & 0 & 0 & 0 & 0 & 1 & 0 & 0 & 0 & 0 & 1 & 0 & 0 & 1 & 0 \\
0 & 1 & 0 & 1 & 0 & 0 & 0 & 0 & 0 & 0 & 0 & 1 & 0 & 0 & 1 \\ \hline
0 & 1 & 0 & 0 & 0 & 1 & 0 & 1 & 0 & 0 & 0 & 0 & 1 & 0 & 0 \\
0 & 0 & 1 & 1 & 0 & 0 & 0 & 0 & 1 & 0 & 0 & 0 & 0 & 1 & 0 \\
1 & 0 & 0 & 0 & 1 & 0 & 1 & 0 & 0 & 0 & 0 & 0 & 0 & 0 & 1 \\ \hline
0 & 1 & 0 & 0 & 1 & 0 & 0 & 0 & 1 & 0 & 1 & 0 & 0 & 0 & 0 \\
0 & 0 & 1 & 0 & 0 & 1 & 1 & 0 & 0 & 0 & 0 & 1 & 0 & 0 & 0 \\
1 & 0 & 0 & 1 & 0 & 0 & 0 & 1 & 0 & 1 & 0 & 0 & 0 & 0 & 0
\end{array}\right).$$

Now we want to project into the field $GF(3) = \{0,1,2\}$. We define

$$\varphi(0) = 0, \quad \varphi(1) = 1, \quad \varphi(\alpha) = 0, \quad \varphi(1+\alpha) = 2.$$

The $GF(3)$-rank of the "transformation matrix"

$$\begin{pmatrix}
1 & 2 & 0 \\
0 & 1 & 2 \\
2 & 0 & 1
\end{pmatrix}$$

is 2. Hence the linear complexity of the sequence

$$(0,1,1,0,1,0,2,2,1,2,0,0,0,2,0,\ldots)$$

is 8.

Part of theorem 3.2.13 is due to Chan and Games [42]. They only determined the complexities for m-sequences (b_i). The interesting thing about this theorem is that it shows how to obtain sequences with a large linear complexity, and which are rather easy to generate. In most applications, the sequence will become binary. Recently, Klapper [109] pointed out that binary sequences $(\varphi(b_i))$ have one disadvantage: There is always a prime p (namely the prime dividing q) such that the sequence $(\varphi(b_i))$ has a small linear complexity considered over a field of characteristic p. So our sequences have a large complexity, which is sometimes useful, but they do not have a large complexity over all fields. Klapper also designed an algorithm how to find the prime p when the sequence $(\varphi(b_i))$ is given. But I expect that the linear complexities of the binary sequences $(\varphi(b_i))$ over fields of characteristic p are larger if we choose a GMW-sequence and not an m-sequence. This is true at least if q is a prime, see Pott [150]. It is therefore good to start with a GMW-sequence rather than an m-sequence. The sequences that we can obtain from Theorem 3.2.13 have the same rank, no matter whether we start from an m-sequence or a GMW-sequence. But regarding Klapper's criticism, the GMW-sequences seem to be better!

3.3 The Waterloo problem

We have seen that relative difference sets are quite rare. Whenever there is a lifting to an abelian relative (m, n, k, λ)-difference set, there is also a lifting to a relative difference set where n is a prime using the standard projection argument. In order to understand why there are not many difference sets which admit liftings to relative difference sets, one should begin with a systematic investigation of liftings "by small primes". In the proof of Proposition 3.2.12 (which will be finished using the results in this section), the case $n = 2$ has been particularly important, and it is exactly this case which will be considered here. Moreover, we will construct another series of relative difference sets with parameters which have not been covered by earlier constructions (see Theorem 2.2.13).

Relative difference sets can be divided naturally into two classes, the splitting and the non-splitting case. We have seen that liftings with $n = 2$ of the trivial (m, m, m)-difference sets are closely related to Hadamard difference sets, liftings of $(m, m-1, m-2)$-difference sets will be investigated in the Sections 5.2 and 6.2. What are the other examples known to us so far? They all come via projections from $((q^d - 1)/(q - 1), q - 1, q^{d-1}, q^{d-2})$-difference sets with q odd. If q is odd and d is even, these examples are non-splitting, if d is odd, they are splitting:

	d even	d odd
q even	?	?
q odd	non-splitting	splitting

In this section, we will give answers to the question marks and prove the non-existence of splitting liftings with q odd and d even.

splitting extensions of complements of the classical Singer difference sets		
	d even	d odd
q even	no	yes
q odd	no	yes

The splitting case is, in my opinion, the more interesting one. In order to explain this, we need the following lemma.

Lemma 3.3.1 (Mullin, Stanton [137]) *Let D be an (m, k, λ)-difference set in G. An $(m, 2, k, \lambda)$-difference set R in $G \times N$ relative to N (which extends D) exists if and only if we can decompose $D = E \cup F$ with $E \cap F = \emptyset$ such that*

$$(E - F)(E - F)^{(-1)} = k \quad in \quad \mathbf{Z}[G]. \tag{3.15}$$

Proof. If $N = \{1, i\}$, then the relative difference set can be written $R = E + Fi$ in $\mathbf{Z}[G \times N]$, $E, F \in \mathbf{Z}[G]$. It is obvious that $E \cup F = D$ and $E \cap F = \emptyset$ since the image of R in $\mathbf{Z}[G]$ is the difference set D. Using the standard equation for the relative difference set R, we see that the group ring elements E and F satisfy the following two equations:

$$EE^{(-1)} + FF^{(-1)} = k + \frac{\lambda}{2}(G - 1), \tag{3.16}$$

$$EF^{(-1)} + FE^{(-1)} = \frac{\lambda}{2}(G - 1). \tag{3.17}$$

These two equations imply (3.15).

Conversely, if (3.15) is true, we combine it with the difference set equation

$$(E + F)(E + F)^{(-1)} = (k - \lambda) + \lambda G \quad in \quad \mathbf{Z}[G]$$

and we obtain (3.16) and (3.17). From these two equations it is straightforward to show that $E + Fi$ is the group ring description of the desired relative difference set. $\square$

We note that this corollary immediately shows that k is a square, say $k = s^2$, and that E and F are sets of respective cardinalities $(s^2 \pm s)/2$: We just have to apply the principal character to (3.15).

Corollary 3.3.2 *If an (m, k, λ)-difference set admits a splitting lifting with $n = 2$, then k has to be a square and λ has to be even.* $\square$

We call the conditions in this corollary the **trivial necessary conditions**. Using Corollary 3.3.2 and some facts from elementary number theory, it is possible to prove the following result:

Result 3.3.3 (Arasu, Jungnickel, Ma, Pott [11]) *Suppose that q is a prime power and d an integer with $d \geq 3$. Then <u>no</u> abelian difference set with parameters*

$$\left(\frac{q^d - 1}{q - 1}, \frac{q^{d-1} - 1}{q - 1}, \frac{q^{d-2} - 1}{q - 1}; q^{d-2} \right)$$

admits a splitting lifting to a relative difference set with $n = 2$.

This is the reason why we consider the complements of the classical parameters. This case is more interesting since the trivial necessary conditions are satisfied if d is odd!

Suppose that D is a cyclic $(2^d-1, 2^{d-1}, 2^{d-2})$-difference set and put $v = 2^d-1$. We define a sequence (a_i)

$$a_i = \begin{cases} +1 & \text{if } (i \text{ modulo } v) \in D \\ -1 & \text{if } (i \text{ modulo } v) \notin D \end{cases}$$

Using the properties of the difference set D, one can check

$$\sum_{i=0}^{2^d-2} (-1)^{a_i+a_{i+s}} = \begin{cases} 2^d - 1 & \text{if } s = 0 \\ -1 & \text{if } s = 1, \ldots, 2^d - 2. \end{cases}$$

We can turn this binary sequence into a ternary sequence (b_i) if D admits a splitting lifting with $n = 2$. Let $E \cup F$ denote the decomposition of D according to Lemma 3.3.1. We define

$$b_i = \begin{cases} +1 & \text{if } (i \text{ modulo } v) \in E \\ -1 & \text{if } (i \text{ modulo } v) \in F \\ 0 & \text{if } (i \text{ modulo } v) \notin D \end{cases}$$

A direct translation of (3.15) yields

$$\sum_{i=0}^{2^d-2} (-1)^{b_i+b_{i+s}} = \begin{cases} 2^d - 1 & \text{if } s = 0 \\ 0 & \text{if } s = 1, \ldots, 2^d - 2. \end{cases}$$

We will discuss these so called "perfect autocorrelation sequences" in Chapter 6.1. The property of the relative difference set to create such a perfect sequence has been the motivation for studying this problem. Several people from the University of Waterloo have been working on the problem, see [28], [137] and [155]. Indeed, at the 6th Southeastern conference on Combinatorics, Graph Theory and Computing in 1975, Collins, Mullin and Schellenberg all gave talks on their independent attacks on this problem. This is the reason why the problem has been sometimes called the **Waterloo problem**. The complete answer to the problem has been given by Arasu, Dillon, Jungnickel and Pott. Let us begin with a positive answer:

Theorem 3.3.4 (Arasu, Dillon, Jungnickel, Pott [8]) *Let q be a prime power, and let d be an odd integer $d \geq 3$. Then there exists a splitting relative difference set R with parameters*

$$\left(\frac{q^d - 1}{q - 1}, 2, q^{d-1}, \frac{q^{d-2}(q-1)}{2} \right)$$

extending the complements of the classical Singer difference sets.

The proof of this theorem uses some facts about quadrics in projective geometries. Everything we need is contained in Games [69], in particular, Games paper contains a generalization of the result about the intersection sizes of quadrics with hyperplanes which we use in the proof of Theorem 3.3.4. The reader might also consult any other text book on projective geometry. Our proof illustrates that (at least some) interesting geometric objects in $PG(d, q)$ have "nice" representations in terms of the Singer difference set.

If q is odd we can also construct splitting difference sets with the desired parameters using projections of the classical affine difference sets. However, this argument fails if q is even. In the following proof, we will present a construction both for the even and odd case. It is not known to me whether the two different constructions in the q odd case yield isomorphic designs.

Proof of 3.3.4. We will construct the decomposition $E \cup F$ (according to Lemma 3.3.1) of the difference set D which consists of the non-zero elements z of trace 0 in the Galois extension $GF(q^d)/GF(q)$. More precisely, the difference set D consists of elements in the quotient group $GF(q^d)^*/GF(q)^*$.

First of all, we show that the following sets are non-degenerate quadrics in $PG(d - 1, q)$:

$$Q_e := \{z \in GF(q^d)^* : \text{trace}(z^{q+1}) = 0\} \quad \text{if } q \text{ is even}$$

and

$$Q_o := \{z \in GF(q^d)^* : \text{trace}(z^2) = 0\} \quad \text{if } q \text{ is odd.}$$

It is obvious that the elements in Q_o satisfy a quadratic equation. The same is true for the elements in Q_e: We choose a normal basis $\{\alpha^{q^i} : i = 0, \ldots, d - 1\}$ of $GF(q^d)/GF(q)$ (see Jungnickel [101] for the definition and construction of normal basis). If $z = \sum a_i \alpha^{q^i}$, we get

$$\text{trace}(z^{q+1}) = \text{trace}(\sum a_i \alpha^{q^i})(\sum a_i \alpha^{q^{i+1}}) =$$
$$= \text{trace}(\sum b_{i,j} \alpha^{q^i} \alpha^{q^j}) = \sum b_{i,j} \text{trace}(\alpha^{q^i} \alpha^{q^j})$$

where the $b_{i,j}$'s are quadratic expressions in the a_i's. In order to prove that the quadrics are non-degenerate, we have to show that there is no point $z \in Q$ with the property

$$x \in Q \Rightarrow (z + x) \in Q \quad \text{for all } x \in Q$$

since in this case each line through z would intersect the quadric either in only one point, or the line would be contained in Q (and Q would be degenerate).

Let us first consider the q <u>even</u> case and assume that $z \in Q_e$ satisfies for all $x \in Q_e$

$$
\begin{aligned}
0 &= \text{trace}(z + x)^{q+1} \\
&= \text{trace}(z^{q+1}) + \text{trace}(x^{q+1}) + \text{trace}(z^q x) + \text{trace}(x^q z) \\
&= \text{trace}(z^q x) + \text{trace}(x^q z).
\end{aligned}
$$

But then $\text{trace}(x(z^q + z^{q^{-1}})) = 0$ for all $x \in Q_e$. In other words: If $z + z^{q^{-1}} \neq 0$, the quadric Q_e would be (for reasons of cardinality) the hyperplane $\{x :$

$\text{trace}(x(z^q + z^{q^{-1}})) = 0\}$. But Q_e is not a hyperplane since $q + 1$ is never a multiplier of a Singer difference set, see Theorem 3.1.1, thus $z^q + z^{q^{-1}}$ has to be 0, equivalently $z^{q^2} = z$ (note that q is even). This shows that $z \in \text{GF}(q)$ since there is no quadratic extension of $\text{GF}(q)$ in $\text{GF}(q^d)$ (since d is odd). But we have $\text{trace}(y) = y$ for elements in $\text{GF}(q)$, therefore $\text{trace}(z^{q+1}) \neq 0$, contradicting $z \in Q_e$.

The case q o̲d̲d̲ is similar: $\text{trace}(z + x)^2 = 2 \cdot \text{trace}(zx)$ (if $\text{trace}(z^2) = \text{trace}(x^2) = 0$) and hence the quadric Q_o would again be a hyperplane, which is absurd since 2 is again not a multiplier.

It is known that there is exactly one tangent hyperplane T_z through each point z of a non-degenerate quadric. The hyperplane T_z is defined via the property that a line through z (in T_z) that meets the quadric Q in a point different from z is already contained in Q. We consider the case q e̲v̲e̲n̲ first: The points x in $T_z \cap Q_e$ must satsify $\text{trace}(z^q x) + \text{trace}(x^q z) = 0$. This holds for $T_z := \{x : \text{trace}(x(z^q + z^{q^{-1}})) = 0\}$. Since $\text{trace}(z^q) = \text{trace}(z^{q^{-1}})$, we have $\text{trace}(z^q + z^{q^{-1}}) = 0$, therefore $\{z^q + z^{q^{-1}} : z \in Q_e\} = \{z : \text{trace}(z) = 0\}$. Note that $T_z \neq T_{z'}$ for $z \neq z'$, hence the set on the lefthand side of the equation above has the same cardinality as the "trace 0"-hyperplane D. Using this difference set D, we can say that the set of tangent hyperplanes is $\{Dz^{-1} : z \in D\}$.

Now we have to use the following observation: Let A and B denote group ring elements (over the integers) corresponding to subsets of a group G, then $|Ax^{-1} \cap B| = (B^{(-1)}A)_x$ (coefficient of x in $B^{(-1)}A$). Moreover, we need the following properties about the intersection of hyperplanes with non-degenerate quadrics: In $\text{PG}(2f, q)$, there are exactly $(q^{2f} - 1)/(q - 1)$ hyperplanes which intersect a non-degenerate quadric Q in $a := (q^{2f-1} - 1)/(q - 1)$ points (these are the tangent hyperplanes) and there are $(q^{2f} \pm q^f)/2$ hyperplanes with an intersection of $a \pm q^{f-1}$ points with Q. We obtain the following equation in $\mathbf{Z}[G]$ (with $d = 2f + 1$):

$$DQ_e^{(-1)} = aG + q^{f-1}(E - F) =: X$$

where E and F denote two disjoint subsets with $E \cup F = G - D$, the complement of the "trace 0"-hyperplane. We will show $(E - F)(E - F)^{(-1)} = q^{2f}$ by computing $XX^{(-1)}$ (here $n = q^{2f-1}$ is the order of the point-hyperplane design):

$$XX^{(-1)} = (n + aG)(n + aG) = n^2 + tG =$$
$$= sG + q^{2f-2}(E - F)(E - F)^{(-1)}$$

for suitable $s, t \in \mathbf{Z}$. We have $t = 2na + a^2(q^{2f+1} - 1)/(q - 1)$ and $s = a^2(q^{2f+1} - 1)/(q - 1) + 2aq^{f-1}q^f$ and therefore $s = t$. This shows that the difference set that describes the complement of the classical point-hyperplane design of $\text{PG}(d-1, q)$ admits a splitting lifting with $n = 2$ if d is odd.

The case q o̲d̲d̲ needs some modification: The tangent hyperplane through $z \in Q_o$ is $\{x : \text{trace}(xz) = 0\} = Dz^{-1}$ where $D = \{x : \text{trace}(x) = 0\}$. We obtain the equation

$$DQ_o^{(-1)} = bG + q^f(B - C),$$

but now $B \cup C = G - Q_o$, and we would get a decomposition of the complement of Q_o. But Q_o is itself a difference set equivalent to D, and this is enough to prove our theorem.   □

Example 3.3.5 Let us illustrate the construction in order to find a $(31, 2, 16, 4)$-difference set. The Singer difference set is

$$D = \{1, 2, 3, 4, 6, 8, 12, 15, 16, 17, 23, 24, 27, 29, 30\}.$$

We have $q = 2$ and $d = 5$. The quadric Q_e is

$$
\begin{aligned}
Q_e \quad &:= \quad \{0 \leq i \leq 30 : 3i \in D\} \\
&= \quad \{1, 2, 4, 5, 8, 9, 10, 11, 13, 16, 18, 20, 21, 22, 26\}.
\end{aligned}
$$

The element $X = DQ_e^{(-1)}$ is (where g generates the cyclic group of order 30)

$$
\begin{aligned}
X \quad &= \quad 7D + 9(g^7 + g^{11} + g^{13} + g^{14} + g^{19} + g^{21} + g^{22} + g^{25} + g^{26} + g^{28}) + \\
&\quad\ + 5(g^0 + g^5 + g^9 + g^{10} + g^{18} + g^{20}),
\end{aligned}
$$

hence we obtain the (cyclic) relative difference set

$$R = \{5, 9, 14, 22, 26, 28, 31, 38, 41, 42, 44, 49, 50, 51, 52, 56\}.$$

We will now show that the condition "d is odd" is also a necessary condition:

Theorem 3.3.6 (Arasu, Dillon, Jungnickel, Pott [8]) *Let q be a prime power and d an even integer, $d \geq 2$. Then no splitting relative difference sets with parameters*

$$\left(\frac{q^d - 1}{q - 1}, 2, q^{d-1}, \frac{q^{d-2}(q - 1)}{2} \right)$$

can exist.

Proof. Let D be a difference set with the Singer parameters in H, and let $E \cup F$ be the decomposition of D according to Lemma 3.3.1. We define $M := E - F$ in $\mathbf{Z}[H]$. We write $d = 2a$ and denote by p the prime satisfying $q = p^\alpha$ for some α. Since the k-value of the underlying difference set D must be a square (Corollary 3.3.2), α is even. Let H be the group which contains D. We may select a subgroup U of H of order $(q^a - 1)/(q - 1)$ and index $w = q^a + 1$ in H. We extend the canonical epimorphism ψ from H onto H/U to $\mathbf{Z}[H]$. We have

$$\psi(M)\psi(M)^{(-1)} = q^{d-1}. \tag{3.18}$$

But p is self-conjugate modulo w, hence we can apply Corollary 1.2.5 to show that the coefficients of $\psi(M)$ are divisible by $q^{(d-1)/2}$. But on the other hand, the coefficients are bounded by $|U|$ since M has only coefficients 0 and ± 1. The easy inequality

$$q^{(d-1)/2} = q^{(2a-1)/2} > (q^a - 1)/(q - 1) = |U| \quad \text{for } q \geq 3$$

implies $\psi(M) = 0$. But this contradicts (3.18) and proves the theorem.   □

Two remarks concerning this non-existence result are in order: First of all, the result holds for all abelian difference sets with the classical parameters and not only for the Singer difference sets. Second, examples of cyclic and non-splitting difference sets with d even can be obtained from the classical Singer difference sets with q odd via projection. If q is even, it is impossible to project onto a difference set with a forbidden subgroup of order 2.

What can we say about liftings of non-classical difference sets? Projections of affine GMW-difference sets yield extensions of the classical GMW-difference sets, hence the GMW-difference sets admit splitting liftings with $n = 2$ if q is odd and d is odd, they have non-splitting liftings if q is odd and d is even. If d is even, they cannot admit splitting liftings.

Problem 9 Is there a lifting construction for the classical GMW-difference sets if q is even?

We run a computer search to find all possible liftings with $n = 2$ of the four inequivalent (and non-isomorphic) cyclic $(121, 81, 54)$-difference sets listed in Baumert [26], see the following table.

A complete list of abelian relative $(121, 2, 81, 27)$-difference sets										
Type A										
0	1	3	9	10	11	16	17	25	27	28
30	31	33	35	37	38	40	47	48	51	55
58	59	62	71	73	74	75	76	80	81	84
86	89	90	91	93	94	99	100	105	106	111
112	114	116	118	120	125	133	141	144	151	155
157	165	167	173	174	177	181	182	185	186	188
190	191	200	203	213	217	219	222	223	224	225
228	229	236	240							
0	1	3	9	23	27	31	32	37	38	44
46	47	52	58	59	62	69	70	71	74	79
80	81	91	93	96	98	100	104	107	111	114
125	131	132	133	137	138	141	146	149	151	154
155	156	157	161	169	172	174	176	177	181	182
185	186	188	194	196	197	203	205	207	210	211
213	215	220	222	223	224	226	227	229	233	236
237	239	240	241							
Type B										
0	4	5	12	15	19	22	23	25	29	31
36	37	45	47	52	53	57	59	62	66	69
74	75	79	80	82	87	88	89	91	93	98

107	108	111	111	122	124	130	134	135	137	141
148	156	159	160	161	163	169	171	177	179	181
182	186	188	191	194	197	198	202	206	207	215
221	222	224	225	226	227	230	233	235	236	237
238	239	240	241							

Type C

0	1	3	4	9	11	12	13	20	27	31
32	33	34	36	37	39	40	46	49	53	55
56	60	62	64	70	74	77	79	81	82	85
91	92	93	94	96	97	99	102	104	107	108
109	111	112	113	117	118	120	131	138	143	146
147	149	151	159	165	168	172	179	180	182	186
187	192	196	197	199	205	209	210	211	221	222
227	231	235	237							

Type D

0	1	3	4	7	9	11	12	19	20	21
23	27	29	32	33	36	38	46	49	52	55
56	57	58	60	62	63	67	69	70	74	77
81	82	83	87	96	97	98	99	100	103	104
108	113	114	115	119	137	138	143	146	147	156
161	165	168	169	171	172	174	180	182	186	187
189	194	196	199	201	207	209	210	215	222	226
231	233	239	241							

0	11	17	19	23	25	26	29	33	38	47
51	52	55	57	58	59	62	68	69	74	75
77	78	80	87	89	98	99	100	114	122	124
125	128	130	133	137	141	142	143	148	153	156
157	161	165	167	169	170	171	174	177	181	182
184	186	187	188	191	194	202	203	204	207	209
215	217	218	222	224	225	226	229	231	233	234
236	239	240	241							

It turns out that the classical Singer difference set (type A) admits two liftings: One of the liftings comes via projection of the affine difference set, the other one via the construction in Theorem 3.2.3. It might be interesting to decide whether the two constructions always yield inequivalent or non-isomorphic relative difference sets.

Chapter 4

Semiregular relative difference sets

Recall that semiregular relative difference sets are extensions of trivial (m, m, m)-difference sets. We have described the connection between $(m, 2, m, m/2)$-difference sets and Hadamard difference sets in Section 2.2. These are the only examples of semiregular relative difference sets in groups which are not p-groups: All known examples with $n \neq 2$ live in p-groups. On the other hand, there is not much evidence that this has to be the case, and it would be nice to get more necessary conditions on the parameters of semiregular examples.

In this chapter, we will restrict ourselves mostly to the case that the group containing the difference set is a p-group: There are a lot of interesting questions even under this restriction on the parameters. We will write the parameters in the form

$$(p^a, p^b, p^a, p^{a-b}) \tag{4.1}$$

where p is always a prime, and a and b are integers with $a \geq b$. The group G is always the group which contains the relative difference set, and N denotes the forbidden subgroup.

The existence question for relative difference sets (4.1) is solved: They always exist as projections of relative $(p^a, p^a, p^a, 1)$-difference sets. If p is odd, we have examples in elementary abelian groups, if $p = 2$, the group is the direct product of $\mathbf{Z}_4$'s (Theorems 2.2.9 and 2.2.10). But there are many more examples of relative difference sets (4.1) if $a \neq b$ which cannot be constructed via projections from the "classical" examples. In the first section, we will give several necessary conditions on the existence of difference sets with parameters (4.1) ("exponent bounds"). Then we will construct some interesting examples which show that some of these necessary conditions are also sufficient.

4.1 Necessary conditions

We begin with an exponent bound for arbitrary abelian $(m, n, m, m/n)$-difference sets. The main idea behind the proof is contained in an old paper by Hoffman [87] where he proved that no cyclic $(n, n, n, 1)$-difference sets can exist if $n > 2$:

Theorem 4.1.1 (Pott [147]) *Let R be an abelian relative $(n\lambda, n, n\lambda, \lambda)$-difference set in G relative to N. Let g be an element in G. Then the order of g divides $n\lambda$ or $n = 2$, $\lambda = 1$ and $G \cong \mathbf{Z}_4$.*

Proof. We consider the coset gN ($\neq N$). Every element in this coset has exactly λ representations as a quotient with elements from R. We multiply all these quotients and get an element

$$S = g^{\lambda n} \cdot \left(\prod_{t \in N} t \right)^{\lambda}.$$

For every element $r \in (xN) \cap R$, there is exactly one element $r' \in (g^{-1}xN) \cap R$ such that $rr'^{-1} \in gN$ provided that $xN \neq gN$, otherwise, there is no such element r'. This shows that every element $f \in R \backslash gN$ occurs exactly twice in an expression $rr'^{-1} \in gN$, once the element f and once its inverse, hence $S = 1$.

If n is odd or λ is even or the Sylow 2-subgroup of N is not cyclic, then it is not difficult to see that $\left(\prod_{t \in N} t \right)^{\lambda} = 1$ and hence the order of g divides $n\lambda$. In the other case, $g^{\lambda n}$ is the unique involution in N, and therefore every element in $G \backslash N$ has order $2n\lambda$. Since n is even and λ is odd, this is only possible if $n = 2$ and $\lambda = 1$ and $G \cong \mathbf{Z}_4$. $\qquad\square$

If we specialize this result to p-groups, we obtain the following exponent bound:

Corollary 4.1.2 *Let R be an abelian relative (p^a, p^b, p^a, p^{a-b})-difference set in G relative to N. Then we have*

$$exp(G) \leq p^a \quad or \quad a = b = 1, \ G \cong \mathbf{Z}_4. \qquad\square$$

Another general exponent bound follows from Turyn's result 1.2.6:

Theorem 4.1.3 *Let R be a relative $(m, n, m, m/n)$-difference set in G relative to N. Suppose that U is a normal subgroup of G such that G/U is abelian and contains a cyclic Sylow p-subgroup. If p is self-conjugate modulo the exponent of G/U and if $p^2 | m$, then p divides $m|U \cap N|/n$.*

Proof. We consider the image $\psi(R)$ of R in G/U under the canonical epimorphism $\psi : G \to G/U$. Since p is self-conjugate modulo the exponent of G/U, we have $\chi(D) \equiv 0 \bmod p$ for all characters of G/U. Using Ma's Lemma (Lemma 1.2.14), we obtain

$$\psi(R) = pX + PY$$

for some X and Y in $\mathbf{Z}[G/U]$ (where P is the unique subgroup of order p in G/U. In particular, the coefficients in $\psi(R)\psi(R)^{(-1)}$ are divisible by p. But we know $\psi(R)\psi(R)^{(-1)}$ from (1.8) which proves the assertion. $\square$

The following two theorems strengthen Corollary 4.1.2 (and therefore they have a much more involved proof).

Theorem 4.1.4 (Ma, Pott [127], Schmidt [157]) *Let R be an abelian difference set in G relative to N with parameters $(p^{2f+1}, p^b, p^{2f+1}, p^{2f+1-b})$ (where p is a prime). If p is odd, we get*

$$exp(G) \le p^{f+1}. \tag{4.2}$$

If $p = 2$, then

$$exp(G) \le 2^{f+2} \tag{4.3}$$

and

$$exp(N) \le 2^{f+1}. \tag{4.4}$$

Moreover, in the case $p = 2$, we have

$$exp(N) \le 2^f \quad if \quad exp(G) > exp(N). \tag{4.5}$$

Proof. Let R be an abelian $(p^{2f+1}, p^b, p^{2f+1}, p^{2f+1-b})$-difference set in G relative to N. Let G_1 be a cyclic group of order p^t where p^t is the exponent of G. We choose an epimorphism $\rho : G \to G_1$ such that $|\rho(N)| = p^n$ where p^n is the exponent of N. Note that this is always possible. We obtain

$$\rho(R)\rho(R^{(-1)}) = p^{4f+2-t}G_1 - p^{2f+1-n}N_1 + p^{2f+1}$$

by (1.2.11) (here $N_1 := \rho(N)$). We have $|\chi(\rho(R))|^2 = p^{2f+1}$ for every character χ on G_1 of order at least p^{t-n+1} since these characters are non-principal on N_1. Let $\alpha = \min\{n, f+1\}$ and h be a generator of G_1. By Proposition 1.2.11, we have

$$\rho(R) = \sum_{m=0}^{\alpha-1} \epsilon_m X_m P_m g_m + P_\alpha Y \tag{4.6}$$

with the notation as in Proposition 1.2.11.

We assume $n > f + 1$, i.e. $\alpha = f + 1$. Let

$$\eta : G_1 \to G_1/P_\alpha \tag{4.7}$$

be the canonical epimorphism and let $G_2 := G_1/P_\alpha$, $N_2 := \eta(N_1)$. Note that $|N_2| \ne 1$ since $n > f + 1$. If p is odd, we get $\eta(\rho(R)) = p^{f+1}\eta(Y)$ and therefore

$$p^{2f+2}\eta(Y)\eta(Y^{(-1)}) = p^{4f+2-t+\alpha}G_2 - p^{2f+1-n+\alpha}N_2 + p^{2f+1}$$

which is impossible. If $p = 2$, then $X_{\alpha-1} = X_f = 1 + h^{2^{t-f-2}}$ and $\eta(X_f P_f) = 2^f(1 + x)$ where x is the unique involution in G_2. We obtain

$$
\begin{aligned}
\eta(\rho(R))\eta(\rho(R^{(-1)})) &= 2^{2f+1}(1 + x) + 2^{2f+2}\eta(Y)\eta(Y^{(-1)}) \\
&\quad + 2^{2f+1} \cdot [\ldots] \\
&= 2^{4f+2-t+\alpha}G_2 - 2^{2f+1-n+f+1}N_2 + 2^{2f+1} \quad (4.8)
\end{aligned}
$$

which shows $n \le f + 1$ and contradicts the assumption $n > f + 1$ (in order to see this, note that the coefficient of every element on the righthand side of (4.8) has to be divisible by 2^{2f+1}). In both cases (p even and p odd), we obtain $\exp(N) \le p^{f+1}$ which proves (4.4).

In the p even case, we can say more. Suppose that $\exp(N) = 2^{f+1}$. We have $\alpha = f + 1$ and $X_f = 1 + h^{2^{t-f-2}}$ (in (4.6)) if $t \ge f + 2$. This is true if $\exp(G) > \exp(N)$. If we project now onto $G_2 = G_1/P_\alpha$, then N_1 is in the kernel of this projection. Hence the image of $\rho(R)$ is a multiple of G_2. On the other hand, we get

$$
\eta(\rho(R)) = 2^f \epsilon_f g_f S_2 + 2^{f+1}Y
$$

(where η is defined as in (4.7), and S_2 is the unique subgroup of order 2 in G_2). This is impossible if Y has integer coefficients. This shows (4.5).

We will now show that the exponent of G is at most 2^{f+2}. Since we know already that (4.4) and (4.5) are true, we may assume $\exp(N) \le 2^f$. In our equation (4.6) we have $\alpha = n \le f$ and $P_\alpha = N_1$. If we apply any non-trivial character χ principal on N_1 to (4.6) we get $\chi(\rho(R)) = \chi(N_1Y) = 0$, if χ is non-principal on N_1 we have $\chi(N_1Y) = 0$, too. This shows $N_1Y = 2^{2f+1-t}G_1$ since $\chi_0(N_1Y) = 2^{2f+1}$. Now we assume $g_0 = 1$ and get

$$
\begin{aligned}
B := \epsilon_0 X_0 P_0 + \epsilon_1 X_1 P_1 g_1 &= 2^{f-1}\epsilon_0(1 + h^{2^{t-2}} - h^{2 \cdot 2^{t-2}} - h^{3 \cdot 2^{t-2}}) + \\
&\quad + 2^{f-2}\epsilon_1 g_1(1 + h^{2^{t-3}} - h^{2 \cdot 2^{t-3}} - h^{3 \cdot 2^{t-3}}).
\end{aligned}
$$

It is easy to see that there is always at least one coefficient in B which is less than or equal to -2^{f-1}. We define

$$
C := \sum_{m=2}^{\alpha-1} \epsilon_m X_m P_m g_m.
$$

The coefficients in C are at most $2^{f-3} + 2^{f-4} + \cdots + 2^{f-n} < 2^{f-2}$, hence there is a coefficient in $\rho(R)$ which is at most

$$
-2^{f-1} + 2^{f-2} + 2^{2f+1-t}.
$$

This number has to be greater than 0, hence $t \le f + 2$ which proves (4.3).

Now let us return to the case that p is odd. Again, we have $\alpha = n$ and $P_\alpha = N_1$ and (4.6) becomes

$$
\rho(R) = \sum_{m=0}^{n-1} \epsilon_m X_m P_m g_m + N_1 Y.
$$

Let η be the canonical epimorphism $G_1 \to G_1/N_1$. We have again $\eta(X_i) = 0$ if $i \leq n - 1$. Hence the image of $\rho(R)$ under η is

$$\eta(\rho(R)) = p^n \eta(Y) = p^{2f+1-t+n} \eta(G_1)$$

(note that $\eta\rho$ projects R onto a group of order p^{t-n} and the forbidden subgroup N is in the kernel of this projection). This shows $N_1 Y = p^{2f+1-t} G_1$, see Lemma 1.1.13. We obtain

$$\rho(R) = \sum_{m=0}^{n-1} \epsilon_m X_m P_m g_m + p^{2f+1-t} G_1.$$

Let

$$g = \begin{cases} h^{p^{t-1}} g_0 & \text{if } \epsilon_0 = -1 \\ h^{cp^{t-1}} g_0 & \text{if } \epsilon_0 = +1 \ (\text{where } c \text{ is a non-square modulo } p). \end{cases}$$

Then we get

$$\begin{aligned} \text{coefficient of } g \text{ in } \rho(R) &= p^{2f+1-t} - p^f \\ &\quad + \text{coefficient of } g \text{ in } \sum_{m=1}^{n-1} \epsilon_m X_m P_m g_m \\ &= p^{2f+1-t} - p^f + \gamma \end{aligned}$$

where $|\gamma| \leq p^{f-1} + p^{f-2} + \ldots + p^{f-n+1} < 2p^{f-1}$. Hence we obtain

$$p^{2f+1-t} \geq p^f - \gamma > p^f - 2p^{f-1}.$$

It follows that $t \leq f + 1$ which proves (4.2). $\qquad \square$

Theorem 4.1.5 (Ma, Pott [127]; Pott [147]) *Let R be an abelian difference set in G relative to N with parameters $(p^{2f}, p^b, p^{2f}, p^{2f-b})$ where p is a prime. Then we have*

$$exp(N) \leq p^f \qquad (4.9)$$

and

$$exp(G) \leq p^f exp(N). \qquad (4.10)$$

Proof. The proof is rather analogous to the proof of Theorem 4.1.4. We project onto a cyclic group G_1 of maximal possible size such that the image N_1 of N is a subgroup whose size p^n is the exponent of N. We obtain

$$\rho(R)\rho(R^{(-1)}) = p^{4f-t} G_1 - p^{2f-n} N_1 + p^{2f}.$$

We put $\alpha = \min\{n, f + 1\}$ and get

$$\rho(R) = \sum_{m=0}^{\alpha-1} \epsilon_m X_m P_m g_m + P_\alpha Y.$$

We assume $n \geq f+1$, i.e. $\alpha = f+1$. Let η denote the canonical epimorphism $G_1 \to G_1/P_{f+1}$. Then

$$\eta(\rho(R)) = \pm p^f \eta(g_f) + p^{f+1} \eta(Y)$$

and we conclude that

$$
\begin{aligned}
\eta(\rho(R))\eta(\rho(R^{(-1)})) &= p^{2f} + p^{2f+1} \cdot [\ldots] \\
&= p^{5f+1-t}\eta(G_1) - p^{3f+1-n}\eta(N_1) + p^{2f}
\end{aligned}
$$

which is impossible if $n \geq f+1$, and which proves (4.9).

The coefficients in $\rho(R)$ are in the interval $[0, p^{2f-t+n}]$ (note that p^{2f-t+n} is the index of $U \cap N$ in U where U is the kernel of ρ). Moreover, we have $\chi(\rho(R)) \equiv 0 \bmod p^f$ for all characters of $\rho(G)$. Then Corollary 1.2.5 shows $p^f \leq p^{2f-t+n}$ which proves (4.10).   □

Proposition 4.1.6 (Kumar, deLauney [113]; Ma, Pott [127]) *Let p be a prime, and let R be a splitting (p^a, p^b, p^a, p^{b-a})-difference set in $H \times N$ relative to N. If H is cyclic, then $a \leq 2$.*

Proof. Suppose that p is odd. If a is odd, we get $a \leq (a+1)/2$, i.e. $a = 1$ (see Theorem 4.1.4). If a is even, we project onto a splitting (p^a, p, p^a, p^{a-1})-difference set and obtain $a \leq 1 + (a/2)$ (see (4.10)) which shows $a = 2$. This argument works also if $p = 2$.

Now suppose that $p = 2$ and a is odd. Then a splitting relative $(2^a, 2, 2^a, 2^{a-1})$-difference set would exist which is impossible (Proposition 2.4.2).   □

Schmidt [157] obtained more necessary conditions on the existence of abelian relative difference sets (4.1) using the exponent, the rank, and the position of the forbidden subgroup N in G. We do not state these rather technical results here but refer to the original paper and Schmidt's doctoral thesis [156]. Let us just mention the following interesting application of his results:

Result 4.1.7 (Schmidt [157]) *Let p be a prime and let R be a relative $(p^{2f}, p^b, p^{2f}, p^{2f-b})$-difference set in an abelian group of exponent p^{f+b}. Then b has to be 1.*

4.2   Examples and characterizations

We summarize some of our constructions in the Sections 2.2 and 2.3 in the following Theorem:

Theorem 4.2.1 *Let $q = p^e$ be a prime power. Then any abelian group of order q^{2n+1} and rank at least $e(n+1)$ contains a relative $(q^{2n}, q, q^{2n}, q^{2n-1})$-difference set. If p is odd, then any abelian group of order q^{2n+2} and rank at least $e(n+2)$*

contains a relative $(q^{2n+1}, q, q^{2n+1}, q^{2n})$-*difference set. If p is even, then any group of rank at least $e(n + 1)$ which contains a $\mathbb{Z}_4^e$ as a subgroup contains a relative* $(q^{2n+1}, q, q^{2n+1}, q^{2n})$-*difference set.*

Proof. If the group has order q^{2n+1}, this is Theorem 2.3.6 with $s = 1$. We get the other examples via the recursive construction in Lemma 2.2.3 using $(q, q, q, 1)$-difference sets. This is the reason why we have to distinguish the even and odd case. □

Corollary 4.2.2 (Davis [50]) *Any abelian group of order p^{2e+b} and rank at least $e + b$ with $b \leq e$ contains a relative* $(p^{2e}, p^b, p^{2e}, p^{2e-b})$-*difference set.*

Proof. We can choose $n = 1$ in the theorem above and then quotient a suitable subgroup. □

Theorem 4.2.1 shows that our exponent bounds are quite good. If p is odd or even, we have relative difference sets with parameters

$$(p^{2f}, p, p^{2f}, p^{2f-1})$$

of exponent p^{f+1}. This shows that Theorem 4.1.5 is best possible if $b = 1$. We have examples with parameters

$$(p^{2f+1}, p, p^{2f+1}, p^{2f})$$

in groups of exponent p^{f+1} if p is odd and exponent 2^{f+2} if $p = 2$. This demonstrates the quality of the bound in Theorem 4.1.4, again for the case $b = 1$. It is quite amazing how little we know if $b > 1$. Some progress in the case $p = 2$ is in a recent paper by Davis and Sehgal [55]. Regarding the bound on the exponent of N, we have examples where equality occurs in (4.9), see the following result. We do not know the quality of the bound on the exponent of N if $a = 2f + 1$ is odd.

Result 4.2.3 (Leung, Ma [117]) *Let s, d, r and t be arbitrary integers with $0 < t \leq d$. We define integers a and b via $s = ar + b$, $0 \leq b < r$. Then there exists a relative*

$$(p^{2sd}, p^t, p^{2sd}, p^{2sd-t})$$

difference set in $(\mathbb{Z}_{p^{a+1}})^{2db} \times (\mathbb{Z}_{p^a})^{2d(r-b)} \times N$ relative to N where N is an arbitrary (even non-abelian) group of order p^t.

Problem 10 Find new constructions for relative (p^a, p^b, p^a, p^{a-b})-difference sets with $b > 1$. Moreover, find new exponent bounds, in particular, if a is even.

If $b = 1$, i.e., if we consider relative (p^a, p, p^a, p^{a-1})-difference sets, the exponent bounds are (with some modifications) also sufficient for the existence of a relative difference set:

Result 4.2.4 (Ma, Schmidt [129]) *Let G be an arbitrary abelian group of order p^{2f+1} with $exp(G) \leq p^{f+1}$. Moreover, let N be an arbitrary subgroup of G of order p. Then G contains a difference set with parameters*

$$(p^{2f}, p, p^{2f}, p^{2f-1})$$

relative to N.

The proof of this result is related to the technique used by Kraemer [111] to construct Hadamard difference set in all abelian groups which satisfy Turyn's exponent bound, see Result 2.1.7 and Theorem 2.4.10. Result 4.2.4 generalizes a theorem of Davis [50].

It is possible to characterize the case of exponent p^{f+1}:

Result 4.2.5 (Ma, Schmidt [129]) *Let $G \cong \langle \alpha \rangle \times H$ be an abelian group of order p^{2f+1} and exponent p^{f+1} (p prime) where the order of α is p^{f+1}. Let N be the subgroup of order p generated by α^{p^f}. Then R is a relative $(p^{2f}, p, p^{2f}, p^{2f-1})$-difference set in G if and only if*

$$R = \dot{\cup}_{i=1}^{p^f} H_i g_i, \quad g_i \in \langle \alpha \rangle$$

and R meets each coset of N exactly once. Here the H_i's are the p^f distinct complements of $\langle \alpha \rangle$ in G.

Let N be a subgroup of order p contained in H (and not contained in a cyclic subgroup of order $exp(G)$). If R is a relative difference set, then it can be written as

$$R = \dot{\cup}_{i=1}^{p^f - p^{f-1}} H_i g_i \ \dot{\cup} \ \langle \alpha^{p^f} \rangle R_1$$

where the H_i are those subgroups which are complements of $\langle \alpha \rangle$ and do not contain N. Moreover, R_1 is a suitable subset of G, and the g_i's are suitable elements from $\langle \alpha \rangle$.

The second part of this theorem is not as nice as the first one: Even if we choose R_1 such that R meets each coset of N exactly once, we cannot say that R is a relative difference set with the desired parameters.

The results in this section have shown that the prime 2 is a "special" prime. The exponent bounds as well as the necessary and sufficient conditions have a different form depending on whether p is even or odd. If we look at the extreme case of $(2^a, 2^a, 2^a, 1)$-difference sets R, it is known that they cannot split, i.e. the forbidden subgroup has no complement. This is easy to see since all the involutions have to be contained in N: If there were an involution outside N, this involution would have at least two difference representations $d - d' = d' - d$ with elements from R. Result 5.4.1 also shows that all the projections of R are non-splitting. On the other hand, splitting examples of relative difference sets in 2-groups are known: Theorem 4.2.1 shows that splitting $(2^{2f}, 2^f, 2^{2f}, 2^f)$-difference sets exist. Note that relative (m, n, k, λ)-difference sets in 2-groups with $m = 2^{2f+1}$ never split (see Proposition 2.4.2). In view of these remarks, the next result is of interest:

Result 4.2.6 (Schmidt [157]) *Abelian splitting difference sets with parameters* $(2^{2f}, 2^{f+1}, 2^{2f}, 2^{f-1})$ *do not exist.*

Chapter 5

Projective planes with quasiregular collineation groups

5.1 Introduction

In this chapter, we will investigate finite projective planes with some "nice" automorphisms. It is widely conjectured that projective planes of order n exist if and only if n is a prime power. We have called this the prime power conjecture, see Problem 1. It seems that we are far away from an answer to this question. Besides the famous Bruck-Ryser Theorem (Corollary 1.1.6) and the computer result $n \neq 10$ by Lam, Thiel and Swierc [115], no restrictions on n are known. As one might expect, the situation changes if there is a group G acting on the plane. In view of the investigations in this monograph, the results obtained by Hughes [89] are of interst, see also Section 3 in Lander [116]. We note that several of these results have been rediscovered from time to time in special cases. The following result can be used quite often in connection with difference set problems. We state it for arbitrary designs and not just planes:

Result 5.1.1 (Hughes [89]) *Let G be an automorphism group of prime order q of a symmetric $(m, k, \lambda; n)$-design. Let f denote the number of fixed points of G and let $w + f$ be the total number of orbits of G. Then the following holds:*

(a) *n is a square or $w + f$ is odd.*

(b) *If a prime p satisfies $p^j \equiv -1 \mod q$, then p divides the squarefree part of n or w is even and f is odd.*

In the next three sections, we will study projective planes admitting quasiregular collineation groups. These investigations are inspired by the classification of quasiregular collineation groups of finite projective planes. We say that an automorphism group acts **quasiregularly** if the stabilizer of each point p fixes all the points in the orbit of p, in other words, the stabilizer of each point is a normal subgroup. Hence any abelian group acts quasiregularly. We want to obtain necessary conditions on the possible orders of projective planes which admit quasiregular collineation groups of certain types. All these efforts have a common aim: We try to get evidence for the prime power conjecture or even prove it under some assumptions on the collineation groups of the planes which are not too strong ("small collineation groups"). Note that the number of point orbits equals the number of block orbits, see Result 1.3.7.

Result 5.1.2 (Dembowski, Piper [61]) *Let G be a quasiregular collineation group acting on a finite projective plane Π of order n. If $|G| > (n^2 + n + 1)/2$, if t denotes the number of point (or line) orbits of G, and if F denotes the fixed substructure of G (i.e., the set of points and lines fixed by all elements in G), then one of the following must hold:*

(a) $|G| = n^2 + n + 1$, $t = 1$, $F = \emptyset$.

(b) $|G| = n^2$, $t = 3$, F *is an incident point-line pair (a, M).*

(c) $|G| = n^2$, $t = n + 2$, F *is either a line and all its points, or its dual.*

(d) $|G| = n^2 - 1$, $t = 3$, F *is a nonincident point-line pair (a, M).*

(e) $|G| = n^2 - \sqrt{n}$, $t = 2$, $F = \emptyset$. *In this case, one of the point orbits is precisely the set of points of a Baer subplane.*

(f) $|G| = n^2 - n$, $t = 5$, F *consists of 2 points, the line joining them and another line through one of the points.*

(g) $|G| = n^2 - 2n + 1$, $t = 7$, F *consists of the vertices and sides of a triangle.*

(h) $|G| = (n - \sqrt{n} + 1)^2$, $t = 2\sqrt{n}$, $F = \emptyset$.

It was proved by Ganley and McFarland [72] that case (h) cannot occur if $n > 4$. It is mentioned in Dembowski and Piper [61] that an example for $n = 4$ exists. Planes of type (c) are translation planes or dual translation planes. In this case, it is known that the group G has to be elementary abelian, and that n has to be a prime power, see Hughes and Piper [91] or Pickert [139]. In other words: For the class of planes of type (c) in Result 5.1.2, the prime power concecture is proved. Case (a) describes $(n^2 + n + 1, n + 1, 1; n)$-difference sets, which we have called planar difference sets. Quite recently, the prime power conjecture for abelian planar difference sets has been verified by a computer search for orders up to $2,000,000$, see Gordon [78]. The only known abelian example in case (e) comes from the plane of order 4. The case (e) corresponds to relative $(m^2 + m + 1, m^2 - m, m^2, 1)$-difference sets (put $n = m^2$), see Ganley and Spence [73]. The example with $n = 4$ is a $(7, 2, 4, 1)$-difference sets which we have described in Section 3.2. It is remarkable that the order of a projective plane admitting an <u>abelian</u> group of type (e) cannot be a prime power unless $n = 4$, see Ganley and Spence [73]. Thus it is conjectured that this case cannot occur. In my opinion, this case is of particular interest since the corresponding relative difference sets would be liftings of the complements of planar difference sets. Note that liftings exist (see Theorem 3.3.4), but we cannot lift that much that the λ-value becomes 1. Hence the question about planes of type (e) is related to the question "how far" we can extend the complement of a planar difference set to a relative difference set. For case (g) in Result 5.1.2, we refer the reader to Kantor [106]: The desarguesian planes are examples of planes of type (g). On the non-existence side, there are several necessary conditions known which rule out the existence of orders of putative planes of type (g) (similar to the case of planar difference sets) which give evidence for the prime power conjecture. The cases (b), (d) and (f) will be studied in the following three sections. In all these cases, we will consider the incidence structure Π' that consists only of the points and lines in the "largest" orbits. The group G acts faithfully on these orbits (see Theorem 13.1 in Hughes and Piper [91], for instance), and hence G is a Singer group of Π'. So one question is what type of incidence structure is Π' and is it possible to reconstruct the plane Π from the residual incidence structure Π'. In all of the cases (b), (d) and (f), this is possible. Thus we will study Π' rather than Π since it is Π' which has a Singer group.

We will use the character theoretic methods which we have developed earlier in this monograph. It turns out that the planes of type (b) and (d) can be described by $(n + 1, n - 1, n, 1)$- and $(n, n, n, 1)$-difference sets (the residual incidence structures are divisible designs with parameters $(n+1, n-1, n, 1)$ and $(n, n, n, 1)$). Thus these relative difference sets are liftings of trivial difference sets! When we investigate these relative difference sets, we can use, of course, all the general results about relative difference sets which we have obtained so far, in particular, we can use the results about the semiregular case in Section 4.1. But we will prove more!

5.2 Affine difference sets

If the plane admits a group of type (d), the incidence structute Π' will be a divisible $(n + 1, n - 1, n, 1)$-design: The points are the points different from a which are not on the line M (where the non-incident point-line pair (a, M) is the fixed structure of G). A point class is the set of points on a line through a, and a line class consists of the lines through a point on M. If we describe the plane by an $(n + 1, n - 1, n, 1)$-difference set in G relative to N, then the cosets of N will describe the point classes. Let us add a point "∞" to the $n^2 - 1$ group elements (which are the points of the divisible design). Furthermore, we define the cosets of the forbidden subgroup N together with ∞ as new blocks. This construction yields an *affine plane* (an affine plane is just a projective plane minus a line and the points on this line, see [91], for instance). This is the reason why $(n+1, n-1, n, 1)$-difference sets are called **affine difference sets** (in slight contrast to the terminology used earlier). We have constructed affine difference sets in Theorem 2.2.12 (put $d = 2$): These are cyclic examples and planes with a cyclic collineation group of type (d) in Result 5.1.2 are called **cyclic affine planes**. This terminology is due to Bose [31] who was the first who investigated this situation. Subsequent and more systematic investigations are Hoffman [87] and recently Jungnickel [99]. As in [99], we will <u>not</u> restrict ourselves to cyclic groups. What is known about these planes: The only known examples arise from desarguesian planes, and it is therefore (sometimes) conjectured that this has to be the case. We know the cyclic examples of Theorem 2.2.12 and also some non-abelian examples that describe desarguesian planes of odd order, see Ganley and Spence [73]. No non-abelian examples seem to be known for even order planes. I do not know whether an analogous result to the "Karzel theorem" holds for affine difference sets: Karzel was able to show that abelian and finite but non-cyclic planar difference sets have to describe non-desarguesian planes (but no examples are known since every known planar difference set describes a desarguesian plane), see Result 1.4.19 in Dembowski [59].

To some extent, the investigation of affine difference sets is analogous to the investigation of planar difference sets. This is true, for instance, for the multiplier theorem for affine difference sets.

Proposition 5.2.1 *Let R be an abelian planar difference set of order n or an abelian affine $(n + 1, n - 1, n, 1)$-difference set. Then every divisor of n is a multiplier.*

Proof. See Theorems 1.3.2 and 1.3.5.   □

But we will also see some differences. It turns out that we can use more geometric arguments in the affine difference set case. The Singer cycle which describes a planar difference set is a nice algebraic object with not much geometric significance. But the forbidden subgroup N of an affine difference set is always a group of (a, M)-homologies: An (a, M)-**homology** is simply a collineation which fixes a non-incident point-line pair (a, M). An (a, M)-**elation** is a collineation fixing an incident point-line pair (we assume that the points on M are fixed!). The line M is called the **axis**, the point a the **center**. We say that a plane is (a, M)-**transitive** for some point-line pair (a, M) if the following holds: Suppose that p and q are two points on a line through a (but $p, q \neq a$, $p, q \notin M$). Then there is a collineation φ such that $\varphi(p) = q$ (it is easy to see that there is at most one such collineation). Hence some of the quasiregular groups in Result 5.1.2 are (p, L)-transitive: For instance, the planes of type (d) are (a, M)-transitive. However, in case (a) we cannot use the Singer group to show that the plane is (p, L)-transitive for some pair (p, L). This might be a reason why we cannot use many geometric arguments to investigate planar difference sets.

Let us consider the case that the plane is (a, M)-transitive for a moment. The points different from a and which are not on M form a divisisble design: The point classes are the points on the lines through a. The design has parameters $(n, n, n, 0, 1)$ if a and M are incident, otherwise it has parameters $(n + 1, n - 1, n, 0, 1)$. The group N of (a, M)-elations or homologies acts regularly on the points of each point class. We say that N acts class-regularly on the divisible design. We define a matrix $\mathbf{H}$ over $\mathbf{Z}[G]$ as follows: Choose a point from each point class and a block from each block class. We label the rows and columns of $\mathbf{H}$ by these elements and define the (p, B)-entry to be the element in N which maps p onto a point on B. Note that this element is uniquely determined since B meets each point class at most once. If a and M are not incident, then there is a point class disjoint from B in which case the corresponding entry in the matrix becomes 0. If (a, M) is an incident point-line pair, we obtain a generalized Hadamard matrix, in the other case, we speak about generalized conference matrices. We refer the reader to Jungnickel [95] for more information about such matrices, see also Proposition 2.2.7.

We want to start our investigation of abelian affine difference sets in G with proofs of the following two results. They are contained in Arasu and Pott [16]:

- The Sylow 2-subgroup of G is cyclic.

- The Sylow 2-subgroup of the multiplier group contains exactly one involution. Hence if the multiplier group is abelian (which is true if G is cyclic), then the Sylow 2-subgroup of the multiplier group is cyclic.

For this purpose, we must investigate polarities of the corresponding planes. For the necessary geometric background, we refer to Hughes and Piper [91], again. A **polarity** φ of a plane Π is an isomorphism of order 2 between Π and its dual. A point p (a line L) is called **absolute** if $p \in \varphi(p)$ (if $\varphi(L) \in L$). We can show that affine difference sets give rise to many polarities. To see this, we have to look more closely at the reconstruction of the plane of order n from the affine difference set of order n. As long as we do not use group ring notation, we write the groups in this section additively. The points of the plane are the $n^2 - 1$ elements of G, a point ∞, and $n + 1$ points $\rho(g)$ in G/N (where ρ denotes the canonical epimorphism from G onto G/N). The lines are the $n^2 - 1$ translates $R + x$ ($x \in G$) where we add to each translate an element of G/N. We do this as follows: There is exactly one coset $N + g$ of N which contains no element of $R + x$. Note that $R + x$ has n elements, there are exactly $n + 1$ cosets of N, and $R + x$ meets each coset at most once. We add the element $\rho(g)$ to $R + x$. Now we must find $n + 2$ further lines. These are the cosets $N + x$ of N to which we add the point ∞ and $\rho(x)$. The last line L_∞ consists of the points in G/N. It is not too difficult to check that this gives a projective plane. It is equally easy to see that G acts via right translation as a quasiregular collineation group on this plane with N as the stabilizer of L_∞. The subgroup N is the subgroup of all possible (∞, L_∞)-homologies.

We are now in a position to construct polarities. Proposition 5.2.2 will be the main tool to prove that the Sylow 2-subgroup of G has to be cyclic. We note that constructions of polarities using quasiregular collineation groups have been previously given by Ganley and Spence [73] and Hall [79].

Proposition 5.2.2 *Let R be an abelian affine difference set in G relative to N, and let x be an element of G. Assume that $N \cap R = \emptyset$ (which we may assume* *without loss of generality). Then*

$$
\varphi_x: \quad
\begin{aligned}
g &\mapsto R + x - g, \\
\rho(g) &\mapsto N + x - g, \\
\infty &\mapsto L_\infty,
\end{aligned}
\qquad
\begin{aligned}
R + g &\mapsto x - g \\
N + g &\mapsto \rho(x - g) \\
L_\infty &\mapsto \infty
\end{aligned}
\qquad (5.1)
$$

is a polarity of the plane corresponding to R.

(By abuse of notation, $N + g$ is a subset of G (hence a set of $n - 1$ points in Π) and also the line determined by $N + g$ in the respective plane. The element $\rho(g)$ denotes an element in G/N and hence just <u>one</u> point in Π on L_∞.)

Proof. First of all note that φ_x is well-defined since $\rho(g) = \rho(h)$ implies $N + x - g = N + x - h$. The map φ_x maps points to lines and lines to points, and it has order 2. We still have to show that φ_x preserves incidence and non-incidence, i.e., a point p is on L if and only if $\varphi_x(L)$ is on $\varphi_x(p)$. Since both the plane Π and its dual are finite planes of the same order, it is enough to check that φ_x preserves incidence. Let $g \in G$ be a point. If $g \in R + s$, say $g = r + s$ and $s = g - r$ for some $r \in R$, then $\varphi_x(R + s) = x - s$, and $\varphi_x(R + s)$ is on $\varphi_x(g)$ which is $R + x - g$. Now let us assume $g \in N + s$, say $g = n + s$ with $n \in N$. We have to show that $\varphi_x(g) = R + x - g$ does not meet the coset $N + x - g$. But this is true since $R \cap N = \emptyset$.

Next, we consider points $\rho(g)$ in G/N. The point $\rho(g)$ is on L_∞, and the point ∞ is on $\varphi_x(\rho(g)) = N + x - g$. Moreover, $\rho(g)$ is on those translates of R which do not meet $N + g$. This shows $\rho(g) = \rho(s)$ if $\rho(g)$ is on $R + s$: We have $R \cap N = \emptyset$, and therefore $R + s$ meets all cosets of N with the exception of $N + s$. Since $\varphi_x(R + s) = x - s$ is an element of $N + x - s = \varphi_x(\rho(s)) = \varphi_x(\rho(g))$, we obtain $\varphi_x(R + s) \in \varphi_x(\rho(g))$.

The assertion is quite obvious if we look at the point ∞ and the lines incident with it.   □

In order to prove the theorem about the cyclic Sylow 2-subgroup of G, we need the following result on the number of absolute points of polarities. For proof, we refer to Hughes and Piper [91], again.

Result 5.2.3 *Let φ be a polarity of a projective plane Π of order n. Then φ has at least $n + 1$ absolute points. If the polarity has <u>exactly</u> $n + 1$ absolute points, then the set of absolute points is*

- *an oval if n is odd,*

- *a line if n is even.*

(Here an **oval** is a set of $n + 1$ points no three on a line.)

The main idea in the proof of Theorem 5.2.5 is to determine the number of absolute points of the polarities φ_x using an "averaging" argument (Proposition 5.2.4). Regarding Theorem 5.2.5, we should mention that the statement about the cyclic Sylow 2-subgroup of G is already contained in a paper by Arasu, Davis, Jungnickel and Pott [7] but only for the case that the order n is a non-square. The proof given here is entirely different since it uses only geometric arguments and does not make use of algebraic number theory. It has the advantage that it covers the square case, too. I do not see how to generalize the proof in [7] to the square case. The n even case has been also studied by Arasu [3].

Proposition 5.2.4 *The number of absolute points of the polarities φ_x as defined in (5.1) is exactly $n + 1$.*

Proof. We use the notation of Proposition 5.2.2. Let $a(\varphi_x)$ denote the number of absolute points of φ_x. Then

$$\sum_{x \in G} a(\varphi_x) \geq (n^2 - 1)(n - 1), \tag{5.2}$$

see Result 5.2.3. We have equality in (5.2) if and only if each of the polarities φ_x has exactly $n+1$ absolute points. We are now going to compute $\sum_{x \in G} a(\varphi_x)$ in a second way. First, we consider the points $g \in G$. A point g is absolute if and only if $2g \in R + x$. Thus the number of absolute points of φ_x contained in G is the product of the number of "squares" in $R + x$ (i.e., elements that can be written as $z = 2y$) times the number of elements in T where T denotes the largest elementary abelian 2-group in G. Note that the map $x \mapsto 2x$ is a group homomorphism with kernel T. The pre-image of an element is just a coset of T, and therefore each square in $R+x$ gives rise to $|T|$ absolute points. Furthermore, an element g is contained in exactly n lines of type $R + x$. This accounts for

$$n \, |G^{(2)}| \, |T| = n \, |G| = (n^2 - 1)n \tag{5.3}$$

absolute points $g \in G$ in the summation on the lefthand side of (5.2) where $G^{(2)}$ denotes the set of squares in G (or the image of G under the map $x \mapsto 2x$). Therefore, $|G^{(2)}| \, |T| = |G|$ which explains (5.3).

A point $\rho(g)$ in G/N is absolute if and only if $2\rho(g) = \rho(x)$, hence $\rho(x)$ has to be a square in G/N. Let T' denote the largest elementary abelian 2-group in G/N. Given $\rho(x)$, there are exactly $|T'|$ choices for $\rho(g)$ such that $2\rho(g) = \rho(x)$ provided that $\rho(x)$ is a square in G/N. If $\rho(x)$ is a square, then there are $n - 1$ choices for x such that $2\rho(g) = \rho(x)$. This gives

$$(n - 1) \, |(G/N)^{(2)}| \, |T'| = (n - 1) \, |G/N| = n^2 - 1 \tag{5.4}$$

absolute points in G/N (again, $(G/N)^{(2)}$ denotes the squares in G/N). If we add the integers in (5.3) and (5.4), we get the total number of absolute points which shows the equality in (5.2). Note that ∞ is never an absolute point of φ_x. □

Theorem 5.2.5 (Arasu, Pott [16]) *Let G be an abelian quasiregular collineation group of type (d) in Result 5.1.2, and let N be the stabilizer of a point orbit of size $n + 1$. Then the Sylow 2-subgroup of G is cyclic.*

Proof. We assume that n is odd (otherwise there is nothing to prove) and denote by R the corresponding affine difference set. Observe that all the involutions of G must be contained in N. Otherwise an involution $t \notin N$ would have two difference representations $t = r - r' = r' - r$ with elements $r, r' \in R$. Now we choose $x \in G$ such that $0 \in R - x$. Then all the involutions and the identity element 0 in G are absolute points of φ_x. We know that the set of absolute points is an oval since n is odd (Result 5.2.3), thus there can be only one involution in N (otherwise the line N would intersect the oval in more than two points). This shows that the Sylow 2-subgroup of G is cyclic.   □

Now we will show that the Sylow 2-subgroup of the multiplier group M must be cyclic if G is cyclic and that M has at most one involution in the (possibly more general) abelian case. In the case of ordinary planar difference sets, the analogous result (for cyclic difference sets) was proved by Ho [86]. The affine statement sounds very similar but the proof is quite different. Ho used the fact that the fixed structure of an involution in the multiplier group is a Baer subplane. As we will see, this is not true in the case of affine difference sets. Here we have a numerical multiplier of order 2 (namely the order n), no matter whether n is a square or not. Therefore, in most cases n cannot be a Baer involution and we can actually show that it is <u>never</u> a Baer involution. (A **Baer involution** is a collineation of order 2 such that the fixed structure is a subplane of order $\sqrt{n}$, in particular, n has to be a square.) Finally, Ho's result seems to be difficult to apply for non-existence proofs since it is hard to find multipliers of even order. But in the affine case it is easy to obtain multipliers of even order. It seems to be a very strong condition that all these multipliers result (after raising to a suitable power) in the same involution $g \mapsto n \cdot g$. We will discuss some of these applications. But first of all we state a (well-known) result about the "type" of involutory collineations of a plane. Again, a proof can be found in Hughes and Piper [91].

Result 5.2.6 *Let Π be a projective plane of order n and φ a collineation of order 2. Then φ is <u>either</u> an elation (in which case n is even) <u>or</u> a homology (n has to be odd) <u>or</u> a Baer involution (n has to be a square).*

Theorem 5.2.7 (Arasu, Pott [16]) *The multiplier group M of an abelian affine difference set contains only one involution, namely the numerical multiplier n.*

Proof. We consider the case of even and odd order separately. We use the same notation as before.
Case (a) *n is even.* In this case, the multiplier n fixes the subgroup N pointwise, the element $\rho(0)$ in G/N and ∞. Furthermore, the line G/N is fixed. Thus the multiplier n is an elation with center $\rho(0)$ and axis N (together with the two extra points $\rho(0)$ and ∞ that we have to add). If ω is an involutory multiplier, then ω fixes the points ∞ and $\rho(0)$ and hence the line G/N. We assert that ω is not a Baer involution. Assume otherwise, then n has to be a square, say $n = m^2$, and ω fixes exactly $m + 1$ points on G/N since G/N is a fixed line,

thus a so called *Baer line*. But the fixed points of ω form a subgroup of G/N (obvious) and hence $m + 1$ divides $|G/N| = m^2 + 1$, a contradiction. This shows that ω is not a Baer involution but an elation with axis N and center $\rho(0)$. The line N is the axis of ω since it contains the fixed points $\rho(0)$, 0 (in G) and ∞, and the only line which contains at least three fixed points of an elation has to be the axis. The multiplier n acts as the automorphism $\rho(g) \mapsto -\rho(g)$ on G/N. We claim that ω acts in the same way which proves that ω is the multiplier n (Note that an elation or a homology is already determined by its center, its axis and the image of only one further element.) We know

$$\omega[\rho(g) + \omega(\rho(g))] = \omega(\rho(g)) + \rho(g), \tag{5.5}$$

hence $\rho(g) + \omega(\rho(g)) = \rho(0)$ since ω has no fixed points besides $\rho(0)$, ∞, and the points in N. This proves $\omega(\rho(g)) = \rho(g) = n\rho(g)$.

Case (b) *n is odd.* Now the multiplier n fixes ∞, $\rho(0)$, the points of N, and the unique involution $\rho(t)$ in G/N (note that the Sylow 2-subgroup of G is cyclic, see Theorem 5.2.5). As in case (a), we can show that no involution ω in the multiplier group is a Baer involution, hence it must be a homology: The axis is N (as in case (a)), and the center is $\rho(t)$ since this is the only involution in G/N, and this involution must be fixed by a multiplier. Using (5.5), we see that $\rho(g) + \omega(\rho(g))$ is a fixed point, hence $\rho(g) + \omega(\rho(g)) = \rho(0)$ or $\rho(t)$. If $\rho(g) + \omega(\rho(g)) = \rho(0)$, we can finish the proof as in case (a). Now assume that

$$\rho(g) + \omega(\rho(g)) = \rho(t) \quad \text{for } \underline{\text{all}} \ \ \rho(g) \neq \rho(0), \rho(t). \tag{5.6}$$

Choose an element $\rho(g)$ such that $\rho(g) + \rho(g) \neq \rho(0), \rho(t)$. This is always possible if $|G/N| > 4$ since the Sylow 2-subgroup of G/N is cyclic. We compute from (5.6)

$$\omega[\rho(g) + \rho(g)] = \rho(t) - \rho(g) - \rho(g)$$

and

$$\omega(\rho(g)) + \omega(\rho(g)) = -\rho(g) - \rho(g),$$

a contradiction. If $|G/N| = 4$, there is a map with property (5.6), but this map is the same as the identity map $\rho(g) \mapsto \rho(g)$, and therefore ω would have too many fixed points on G/N. □

Now we leave the investigation of the Sylow 2-subgroups of M and G and try to find numerical restrictions on the orders of affine difference sets. We will use the results developed so far to obtain some information about the "geometry" of multipliers. We will classify those multipliers which are planar. We call a collineation or a multiplier **planar** if its fixed stucture is a subplane. Using the characterization in Theorem 5.2.8, we will obtain strong numerical conditions on the orders of affine difference sets. Some notation: If φ is an automorphism of the group G, we denote by $\text{fix}(\varphi, G)$ the number of fixed points of φ in G, i.e., the number of points with $\varphi(x) = x$.

Theorem 5.2.8 (Arasu, Pott [16]) *Suppose that φ is a multiplier of a (not necessarily abelian) affine difference set R in G relative to the normal subgroup N. Note that φ is also an automorphism of G/N since φ fixes N setwise. Then*

$$\varphi \text{ is a planar multiplier if and only if } \text{fix}(\varphi, G/N) \geq 3. \tag{5.7}$$

If φ is planar, then

$$\text{fix}(\varphi, G/N) = \text{fix}(\varphi, N) + 2 \tag{5.8}$$

and

$$\text{fix}(\varphi, G) = (\text{fix}(\varphi, G/N) - 1)^2 - 1. \tag{5.9}$$

Proof. The multiplier φ fixes the points ∞, 0 and $\rho(0)$, and it fixes the line G/N. It is enough to show that at least three lines through ∞, thus lines of the type $(N + x) \cup \{\infty, \rho(x)\}$, are fixed by φ. This yields a quadrangle, and therefore the fixed structure is a proper projective plane and not a degenerate plane: Two fixed points are joined by a fixed line and two fixed lines intersect in a fixed point, hence the fixed structure is a projective plane if it contains a quadrangle, see Section 1.1. To find these three lines through ∞, note that the line $(N + x) \cup \{\infty, \rho(x)\}$ is fixed if and only if $\varphi(\rho(x)) = \rho(x)$. This proves (5.7). Furthermore, $\text{fix}(\varphi, G/N)$ is the number of fixed points on a line of the subplane, and $m := \text{fix}(\varphi, G/N) - 1$ is the order of this subplane. The number $\text{fix}(\varphi, G/N)$ has to be $\text{fix}(\varphi, N) + 2$ since the line N has $\text{fix}(\varphi, N)$ fixed points plus ∞ and $\rho(0)$. Observe that the subplane of Π fixed by φ contains $\text{fix}(\varphi, G)$ points different from ∞ which are not in G/N. This number must be $m^2 + m + 1 - (m + 1) - 1 = m^2 - 1$, and thus (5.9) follows from (5.8). $\qquad\square$

For cyclic groups and numerical multipliers t, we have $\text{fix}(t, G) = (t - 1, |G|)$. Then part of our theorem reduces to Theorem 3.2 in Hoffman [87].

Theorem 5.2.9 (Arasu, Pott [16]) *Let R be an abelian affine difference set of order n in G relative to N. Let φ be any multiplier (not necessarily numerical). Then the following holds:*

(a) *If φ considered as an automorphism of G/N is the identity, then φ is the identity on G.*

(b) *If $\varphi|N = id$, then $\varphi = id$ or φ is the multiplier n.*

Proof. The first statement follows from our characterization of planar multipliers (Theorem 5.2.8): The multiplier φ has to be planar since it has $n + 1$ (≥ 3) fixed points. But φ fixes the line G/N pointwise, and hence the fixed structure cannot be a proper subplane, i.e., φ has to be the identity.

Now assume that $\varphi|N = id$. Then φ fixes the points on N, the point ∞, and $\rho(0)$ in G/N, hence it fixes the line determined by N pointwise. Hence φ is the identity or an elation or a homology. Assume that φ is not the identity. The center of φ is the intersection of two fixed lines different from N. One of the fixed lines is G/N, hence the center is not a point in G but a point on G/N, and φ fixes no points in G which are not in N. We can actually determine the center: If n is even, then φ is an elation and the center is $\rho(0)$. If n is odd, then φ is a homology where the center is the unique involution in G/N. Note that this is the same fixed structure as those of the multiplier n.

By Theorem 1.3.8, we may assume that the affine difference set is fixed by all multipliers, i.e., if $x \in R$, then $nx \in R$ (since n is a multiplier), $\varphi(x) \in R$, and $n\varphi(x) \in R$. If x is not an element of N, then $x \neq \varphi(x)$ and $nx \neq \varphi(nx)$ since φ and n have no fixed points in $G \setminus N$. We claim that

$$n\varphi(x) - x = nx - \varphi(x). \tag{5.10}$$

To see this, note that $n\varphi(x) - nx - x + \varphi(x) = (\varphi - id)((n+1)x)$. The element $(n+1)x$ is an element of N since $|G/N| = n+1$, therefore $\varphi((n+1)x) = (n+1)x$ since we have assumed $\varphi|N = id$. Equation (5.10) gives two different representations of a non-zero element as a difference with elements from R unless $nx = \varphi(x)$, i.e., unless φ is the multiplier n: As in the proof of Theorem 5.2.7, it is enough to check $\varphi(x) = nx$ for just one non-fixed point x since φ and the multiplier n have the same fixed structure. $\square$

We can translate Theorems 5.2.8 and 5.2.9 into a necessary condition on possible orders of affine difference sets:

Corollary 5.2.10 *Let n be the order of an abelian affine difference set in G relative to N and let t be a divisor of n. Then t has to satisfy the following conditions:*

(a) *The order of t modulo the exponent of G/N is the same as the order modulo the exponent of G.*

(b) *If $t^f \equiv 1 \mod exp(N)$, then $t^f \equiv 1$ or $n \mod exp(G)$.*

(c) *If $t^{2f} \equiv 1 \mod exp(G/N)$, then $t^f \equiv 1$ or $n \mod exp(G)$.*

Proof. First of all we note that t is a multiplier of R.
Case(a). We apply Theorem 5.2.8 (a): If f is the order of t modulo the exponent of G/N, then t^f has to be the identity on G and $t^f \equiv 1 \mod exp(G)$. It is obvious that the order modulo the exponent of G cannot be smaller than the order modulo the exponent of N.
Case (b). This is immediate from Theorem 5.2.9 (b).
Case (c). Case (a) shows that $t^{2f} \equiv 1 \mod exp(G)$, hence the multiplier t^f is the identity or a multiplier of order 2. But the only multiplier of order 2 is the multiplier n (Theorem 5.2.5). $\square$

We note that the conditions in Corollary 5.2.10 are similar to some of the conditions given by Jungnickel and Pott [102]. In that paper it has been checked that the order n of an abelian affine difference set has to be a prime power if $n \leq 10,000$. This extends a result of Ko and Ray-Chaudhuri [110] who considered only cyclic affine planes in the range up to 5,000. The main result that they used is the following:

Result 5.2.11 (Ko, Ray-Chaudhuri [110]) *Let R be a cyclic affine plane of order n and let p be a prime divisor of n.*

(a) *If $p^j \equiv 1 \mod (n+1)$, then $p^j \equiv 1 \mod (n^2 - 1)$.*

(b) *If $p^j \equiv n \mod (n+1)$, then $p^j \equiv n \mod (n^2 - 1)$.*

(c) *If $p^j \equiv 1 \mod (n-1)$, then $p^j \equiv 1$ or $n \mod (n^2 - 1)$.*

If t is a prime in Corollary 5.2.10 and G is cyclic, then Result 5.2.11 follows from Corollary 5.2.10. However, the original proof in Ko and Ray-Chaudhuri [110] seems to work only for cyclic groups, at least I do not see how to extend it to the abelian case. Furthermore, they did not use the geometric language as we did in the proofs above. In my opinion, the geometric interpretation of the multipliers is the best approach to prove a result like 5.2.11.

I did a computer search for the orders n of cyclic affine planes up to 10,000 using Result 5.2.11. Only the following non-prime powers survive:

$$33, 158, 428, 437, 513, 611, 843, 1016, 1184, 1328, 1347, 1405, 1472, 2252, 2427,$$
$$2669, 2763, 3272, 4304, 4568, 5699, 6080, 6662, 6998, 8935, 9068.$$

In the range up to 5,000 these are the numbers Ko and Ray-Chaudhuri couldn't rule out either using Result 5.2.11. Erroneously, they missed the number 1347 which satisfies the necessary conditions of Result 5.2.11: We factor $1347 = 3 \cdot 449$. The order of 3 and 449 modulo 1348 is 168, i.e. $j = 168$ in 5.2.11, and $3^{168} \equiv 1 \mod (1347^2 - 1)$ as well as $449^{168} \equiv 1 \mod (1347^2 - 1)$, no contradiction. The equations $3^j \equiv 1347 \mod 1348$ and $449^j \equiv 1347 \mod 1348$ have no solutions. The order of 3 and 449 modulo 1346 is 168, again. Since $3^{168} \equiv 1 \mod (1347^2 - 1)$ and as $449^{168} \equiv 1 \mod (1347^2 - 1)$, we obtain no contradiction. However, we can rule out the order 1347 using Corollary 5.2.10 (c): We have $3^{168} \equiv 1 \mod (1347^2 - 1)$ but 3^{84} is neither congruent 1 nor 1347 modulo $(1347^2 - 1)$. This example shows that our Theorems are stronger than Result 5.2.11. We have used Corollary 5.2.10 to check the values listed above. The only survivors are

$$33, 513, 843, 1328, 2427, 2763, 3272, 4568, 6080.$$

Of course, it is possible to rule out these orders using other tests, see [102]. But we want to stress that purely geometric reasoning is enough to prove the prime power conjecture for cyclic affine planes of order n with $n \leq 10,000$ with only 9 exceptions.

Now we are going to consider more necessary conditions on the orders of putative affine difference sets. Of course, we can use the Mann test for divisible difference sets (Theorem 2.4.1). Another "type" of theorem is the following result.

Result 5.2.12 *Let R be an abelian affine difference set of order n. If $n \equiv 0 \bmod 2$, then $n = 2, 4$ or $n \equiv 0 \bmod 8$. If $n \equiv 0 \bmod 3$, then $n = 3$ or $n \equiv 0 \bmod 9$.*

For proofs of this result, we refer to Arasu and Jungnickel [10] and Pott [144]. The corresponding result is true for planar difference sets. The proofs for the case that 3 divides n are analogous in the affine and the planar case, see Wilbrink [175]. There are several proofs for the conclusion that 4 has to divide n if n is even and $n \neq 2$, some of them (Jungnickel [96], Pott [144]) are analogous to the planar case (which is due to Jungnickel and Vedder [104] and independently Wilbrink [175]), some of them (Hughes [90]) are not. The proof that n is divisible by 8 if n is even ($n \neq 2, 4$) is very different in the planar [104] and affine [10] case. We will use the method in [10] to extend Result 5.2.12. As a byproduct, we will obtain an easy proof of Result 5.2.12 in the case that n is even.

Theorem 5.2.13 (Arasu, Pott [17]) *Let R be an <u>abelian</u> affine difference set of order $n \equiv 8 \bmod 16$. Then $n - 1$ must be a prime power. If R is a <u>cyclic</u> difference set, then $n - 1$ must be a prime.*

Before we are going to prove this theorem, we want to explain the technique used by Arasu and Jungnickel in [10] to prove "$8|n$" in 5.2.12. Our approach follows Jungnickel [99]. Let R be an abelian affine difference set of even order n in G relative to N. Assume that R is fixed under all multipliers. The group G splits as $G \cong H \times N$ where $|H| = n + 1$ (since n is even). In what follows, we will use group ring notation, and therefore we will switch to the multiplicative notation. We apply the canonical epimorphism $\rho(G) \to N$ to the group ring element R and obtain an element $\rho(R)$ satisfying

$$\rho(R) = n + nN \quad \text{in } \mathbf{Z}[N].$$

Note that the coefficients of $\rho(R)$ are the intersection numbers $|R \cap Hx|$, $x \in N$. Since n is a multiplier that acts as the identity on N, we have the following implication: If $y \in R \cap Hx$, then $y^n \in R \cap Hx$. Thus each coefficient of $\rho(R)$ is divisible by 2 unless $y \in R \cap Hx$ is an element of N. But the intersection of R with N has cardinality at most 1 (standard argument) and hence $|R \cap N| = 0$ since $|R|$ is even. This shows that $P := \rho(R)/2$ is an element in $\mathbf{Z}[N]$. We get

$$PP^{(-1)} = \frac{n}{4} + \frac{n}{4}N. \tag{5.11}$$

Hence n is divisible by 4 if N is not the trivial subgroup, i.e. if $n \neq 2$. Let us write

$$P := \sum_{g \in N} a_g g.$$

We have

$$\sum_{g \in N} a_g = \frac{n}{2} \quad \text{and} \quad \left(\sum_{g \in N} a_g \right)^2 = \frac{n}{2},$$

hence $a_g = 0$ or 1 and P is an $(n-1, n/2, n/4)$-difference set in N. Difference sets
with these parameters or the complementary parameters $(n-1, n/2-1, n/4-1)$
are the Paley-Hadamard difference sets. The existence of these Paley-Hadamard
difference sets was established by Arasu and Jungnickel in [10] using more involved arguments. They finished their proof as follows: The integer 2 is a
multiplier of the affine difference set of even order n, hence a multiplier of the
Paley-Hadamard difference set of order $n/4$. But 2 is only a multiplier of a difference set if 2 divides the order or the difference set is trivial, see Pott [142], for
instance, hence $n \equiv 4 \mod 8$ implies $n = 4$ (in which case the Paley-Hadamard
difference set is trivial with parameters $(3, 2, 1)$).

Proof of Theorem 5.2.13. The order of the Paley-Hadamard difference set
constructed above is congruent 2 modulo 4 if $n \equiv 8 \mod 16$. Arasu [2] proved
that the complement P' of P satisfies $P + P' = 1 + N$ in GF(2)$[N]$ if the order
of P is $\equiv 2 \mod 4$, see also Proposition 6.3.6. Luckily, Camion and Mann [40]
studied exactly these difference sets (and they called them antisymmetric, see
Theorem 2.3.11). Result 5.2.14 finishes the proof of Theorem 5.2.13.   □

Result 5.2.14 (Camion, Mann [40]) *Let D be an abelian (m, k, λ)-difference
set in G which satisfies $D + D^{(-1)} + 1 = G$ in $\mathbf{Z}[G]$. Then m is a prime power
congruent 3 modulo 4. If G is cyclic then m is a prime and D consists of the
non-zero squares or the non-squares in GF(m).*

For cyclic affine planes of order $n \equiv 8 \mod 16$, our theorem together with
Result 5.2.14 implies that the Paley-Hadamard difference set P in N has to be
the set of squares or non-squares in GF($n-1$) where $n-1$ is a prime, i.e., they are
the classical Paley difference sets. In general, the set of squares provide examples
of antisymmetric difference sets and these are the only <u>known</u> examples. (We
refer to Chen, Xiang and Sehgal [44] for a recent investigation of antisymmetric
difference sets). One may ask the question what type of difference set is P if
the affine difference set is the classical cyclic one of order 2^t. The following
proposition answers this question:

Proposition 5.2.15 (Arasu, Pott [17]) *Let R be a classical cyclic affine $(q+
1, q-1, q, 1)$-difference set with $q = 2^t$. Then the Paley-Hadamard difference set
P in the forbidden subgroup constructed above is the complement of the classical
$(2^t - 1, 2^{t-1} - 1, 2^{t-2} - 1)$-Singer difference set.*

Proof. The classical cyclic affine difference set R of order q is a subset of
GF(q^2)*. It consists of the elements ω such that trace(ω) = 1 or any other
non-zero element, but we may assume "trace = 1" w.l.o.g. Here we consider,
of course, the trace function of the quadratic extension GF(q^2)/GF(q). The
forbidden subgroup N is simply GF(q)*. The map $\omega \mapsto \omega^{2^{t-1}(2^t+1)}$ is an epimorphism GF(2^{2t})$^* \rightarrow$ GF(2^t)* which is the identity on N, and whose kernel is the complement of N in GF(2^{2t})*: The element ω^{2^t+1} generates N and

$(\omega^{2^t}+1)^{2^{t-1}(2^t+1)} = \omega^{(2^{2t}+2^{t+1}+1)2^{t-1}} = \omega^{(2^{t+1}+2)2^{t-1}} = \omega^{2^t+1}$. The element ω^{2^t-1} generates the complement of N and $(\omega^{2^t}-1)^{2^{t-1}(2^t+1)} = \omega^{(2^{2t}-1)2^{t-1}} = 1$.

If $\omega + \omega^{2^t} = 1$ (i.e., if the trace of ω is 1), then the trace of $\omega^{2^{t-1}(2^t+1)}$ (now with respect to $\mathrm{GF}(2^t)/\mathrm{GF}(2)$) is

$$\mathrm{trace}_{t/1}(\omega^{2^{2t-1}}\omega^{2^{t-1}}) = \mathrm{trace}_{t/1}(\omega\omega^{2^t}) = \mathrm{trace}_{t/1}(\omega + \omega^2) = \omega + \omega^{2^t} = 1$$

(note that $\omega + \omega^2$ but not ω is an element of $\mathrm{GF}(2^t)$). This shows that the difference set P consists of the elements of trace 1 in $\mathrm{GF}(2^t)$, which is the complement of the classical Singer difference set, see Theorem 2.1.1. $\qquad\square$

In order to prove that cyclic affine difference sets of order $n \equiv 8$ mod 16 with $n \neq 8$ cannot exist (which is presumably true!), it is "enough" to show that the set of squares (or non-squares) together with 0 cannot occur as a Paley-Hadamard difference set P in N. We note that for $n = 8$ (i.e. $|N| = 7$) the Paley-Hadamard and the Singer difference sets coincide. But this is the only case when this happens. In some sense, Paley and Singer difference sets with parameters $(2^t - 1, 2^{t-1}, 2^{t-2})$ are as different from each other as possible. The codes of the complementary Singer difference sets have the smallest possible $\mathrm{GF}(2)$-rank (which is t by Corollary 3.1.6), whereas the $\mathrm{GF}(2)$-rank of the Paley difference sets is as large as possible (it is $2^{t-1} - 1$, see Theorem 6.3.2). So it might be possible to prove that the case $n \equiv 8$ mod 16 is possible only if $n = 8$ using the codes of the "underlying" Paley-Hadamard difference sets. Moreover, I hope that the Paley-Hadamard difference set P turns out to be useful for more non-existence results on affine difference sets if n is even and not only if $n \equiv 8$ mod 16.

It is crucial for our construction of P that the group G splits, i.e., that the forbidden subgroup appears as a direct factor. Thus our construction cannot work if n is odd. In this case, the group can be written $G \cong H' \times N' \times S$ where S is the Sylow 2-subgroup of G, and N' is contained in N, and the index of N' in N is a power of 2. Since the Sylow 2-subgroup of G is cyclic (Theorem 5.2.5), the group S contains a unique subgroup S' of index 2. In what follows, we will describe a construction that produces from an abelian affine difference set an interesting "difference family" in a group isomorphic to N'. We have to assume that the order n of the affine difference set R is congruent 3 modulo 4. Then the index of N' in N is exactly 2. Assume that R is fixed by the multiplier n, and that the identity element $1 \in G$ is an element of R (we write the group multiplicatively). We can assume this without loss of generality: If $g \in R$, then $g^n \in R$ and we have $g^n = g$ if and only if $g \in N$. Therefore, the elements in R come in pairs g and g^n with $g \neq g^n$ unless $g \in N$ since there is only one subgroup of order $n - 1$ in G. Since the number of elements in R is odd, there must be at least (and hence exactly one) element in $R \cap N$, say d. But Rd^{-1} is again fixed by the multiplier n (since $d^n = d$) and $1 \in Rd^{-1}$ which shows that we may indeed assume that $1 \in R$.

Now we factor out the subgroup $H' \times S'$ ($=: H$) of order $n + 1$ and denote the canonical epimorphism $G \to G/H$ by ρ. Note that H is not a complement of N but $|H \cap N| = 2$ (the unique subgroup of G of order 2). We obtain from $RR^{(-1)} = n + G - N$ the equation

$$\rho(R)\rho(R)^{(-1)} = n + (n+1)\rho(G) - 2\rho(N) \quad \text{in } \mathbf{Z}[\rho(G)] \qquad (5.12)$$

where $\rho(G) \cong N$ and $\rho(N)$ has index 2 in $\rho(G)$. The coefficients of $\rho(R)$ are the intersection numbers $|R \cap Hx|$. If $g \in R \cap Hx$, then $g^n \in R \cap Hx$ since $|G/H| = n - 1$ and R is fixed by the multiplier n. Therefore, the coefficients of $\rho(R)$ are divisible by 2 (with the exception of the coefficient of the identity) since $g^n \neq g$ for non-identity elements.

We write

$$\rho(R) + 1 = \sum_{g \in \rho(G)} a_g g$$

and obtain from (5.12) the two equations

$$\sum_{g \in \rho(G)} a_g = n + 1$$

and

$$\sum_{g \in \rho(G)} (a_g)^2 = 2n + 2|R \cap H|$$

(look at the coefficient of 1 in $(\rho(R) + 1)(\rho(R)^{(-1)} + 1)$). The coefficient of the identity in $\rho(R)$ is 1 since $|R \cap H| = 1$: We have $1 \in R$, and if $g \in R \cap H$ $(g \neq 1)$, then $g^n \in R \cap H$ with $g^n = g^{-1}$. Then we get two representations $1g^{-1} = g^{-1}1$ of an element in G using elements of R which is impossible. Hence we have

$$\sum_{g \in \rho(G)} a_g(a_g - 2) = 0$$

and therefore the elements a_g are 0 and 2 (note that they are even!). We define $P := (\rho(R) + 1)/2$ which is an element in $\mathbf{Z}[\rho(G)]$ with 0/1-coefficients. Let us compute $PP^{(-1)}$:

$$
\begin{aligned}
PP^{(-1)} &= \frac{(\rho(R) + 1)(\rho(R)^{(-1)} + 1)}{4} \\
&= \frac{n + (n+1)\rho(G) - 2\rho(N) + \rho(R) + \rho(R)^{(-1)} + 1}{4} \\
&= \frac{(n-1) - 2\rho(N) + 2P + 2P^{(-1)} + (n+1)\rho(G)}{4} \\
&= \frac{(n-1)/2 - \rho(N) + P + P^{(-1)}}{2} + \frac{n+1}{4}\rho(G).
\end{aligned}
$$

Let us write $P := P_1 + P_2 t$ where t denotes the unique involution in $\rho(G)$. The crucial observation in our derivation is that the coefficients of $(n-1)/2 - \rho(N) + P + P^{(-1)}$ have to be even (recall that we assume $n \equiv 3 \bmod 4$). This forces $P_1 + P_1^{(-1)} = 1 + \rho(N)$ and $P_2 = P_2^{(-1)}$. We obtain for the sizes of P_1 and P_2

$$|P_1| = |P_2| = (n+1)/4. \qquad (5.13)$$

The equation for $PP^{(-1)}$ becomes

$$
\begin{aligned}
(P_1 + P_2 t)(P_1 + P_2 t)^{(-1)} &= P_1 P_1^{(-1)} + P_2 P_2^{(-1)} + Xt \\
&= \frac{n+1}{4} + \frac{n+1}{4}\rho(N) + Xt
\end{aligned}
$$

where X is some element in $\mathbf{Z}[\rho(N)]$ which is of no interest to us. We obtain

$$
P_1 P_1^{(-1)} + P_2 P_2^{(-1)} = \frac{n+1}{4} + \frac{n+1}{4}\rho(N). \tag{5.14}
$$

As divisible difference sets describe divisible designs, two subsets A and B in a group H of order v describe a (v, k, λ)-design if $|A| = |B| = k$ and $AA^{(-1)} + BB^{(-1)} = (2k - \lambda) + \lambda H$. The blocks of the design are the translates of A and of B, hence the design is definitely not symmetric! In our special case, we call the two subsets a (v, k, λ)-**difference family** with two start blocks. If we replace A and B by there complements, we get a difference family again. Of course, this generalizes to difference families with more than just one start block, we refer the reader to Beth, Jungnickel and Lenz [29]. Some authors call these objects **supplementary difference sets**. In case of two start blocks, the designs have exactly $b = 2v$ blocks. The equations (5.13) and (5.14) show the following:

Theorem 5.2.16 *Let R be an abelian affine difference set of order $n \equiv 3 \bmod 4$ relative to N. Then there exists an*

$$
\left(\frac{n-1}{2}, \frac{n+1}{4}, \frac{n+1}{4}\right)\text{-difference family}
$$

with two start blocks P_1 and P_2 in N'. Here N' is the unique subgroup of index 2 in N. Moreover,

$$
\begin{aligned}
P_1 + P_1^{(-1)} &= 1 + N', &(5.15)\\
P_2 &= P_2^{(-1)} &(5.16)
\end{aligned}
$$

holds in $\mathbf{Z}[N']$. □

It is straightforward to check that the complements of P_1 and P_2 form a difference family P_1', P_2' with parameters

$$
\left(\frac{n-1}{2}, \frac{n-3}{4}, \frac{n-7}{4}\right). \tag{5.17}
$$

Again, the start blocks satisfy

$$
P_1' + P_1'^{(-1)} = N' - 1 \tag{5.18}
$$

and (5.16) in $\mathbf{Z}[N']$.

Difference families with these parameters are known, not only for prime powers, see Takeuchi [166]. The situation becomes different if we impose conditions

(5.16) and (5.18). These two conditions seem to be very strong (although I cannot use them for non-existence proofs so far): One start block has to be symmetric, the other one antisymmetric! Let us call this property (S-AS). Examples of difference families with property (S-AS) are the **classical Szekeres difference sets** (which are of course not difference sets in our sense and should be better called difference families). One can obtain these as a *derivation* of the Paley difference sets in GF(q) where q is a prime power $\equiv 3 \bmod 4$: Take the set of squares in GF(q) (the Paley difference set). The start blocks of the difference family are $D \cap (D + 1)$ and $D \cap (D - 1)$ considered as subsets of the cyclic multiplicative group of squares.

Yamada [179] has shown that the difference families P_1', P_2' with parameters (5.17) that we can construct from the classical cyclic affine planes are the classical Szekeres difference sets. Other examples of difference families with property (S-AS) and parameters (5.17) have been constructed by Arasu [4]: If $q \equiv 5 \bmod 8$ is a prime power, then $C_0 \cup C_1$ ($= P_1'$) and $C_0 \cup C_2$ ($= P_2'$) is a difference family with property (S-AS). Here C_0 is the set of fourth powers in GF(q)*, and the C_i's are the cyclotomic classes $C_0 \omega^i$ where ω is a primitive element of GF(q). Note that these examples show that $n = 2q + 1$ is not necessarily a prime power (take $q = 37$, $n = 75$), in particular, the classical Szekeres difference sets are <u>not</u> the only difference families with property (S-AS). But I still believe that property (S-AS) should give numerical restrictions on the orders of affine difference sets.

We have defined the classical Szekeres difference sets, and therefore the question arises what the Szekeres difference sets are in general. They are difference families with parameters (5.17) where only one of the start blocks is symmetric (property (5.16)). Szekeres difference sets with the additional property (5.18) are useful ingredients in the construction of Hadamard matrices, see Wallis, Street and Wallis [171]. In this context, the construction of Arasu [4] is already contained in [171, page 304], but only for the special cases $q = 5, 13, 19$ and 53.

5.3   Direct product difference sets

In this section, we consider projective planes Π of order n admitting a quasiregular collineation group G of type (f) in Result 5.1.2. It seems that this type of plane is not yet studied as extensively as the cases (b) and (d) discussed in the previous and in the next section. One reason might be that the residual incidence structure Π' consisting of points and lines in the "big" orbits under the quasiregular collineation group is not a divisible design as in the cases (b) and (d). Now we have two kinds of point classes. Therefore, in the difference set representation we will have two exceptional subgroups. Under the assumption that the group is abelian, we will prove the following:

- If n is even, then $n = 2^e$ and the Sylow 2-subgroup of G is elementary abelian.

- If n is odd, then the Sylow 2-subgroup of G is cyclic.

- If n is not a square, then n is a prime power.

- If n is a prime, then the plane is desarguesian.

The first of these results was already proved by Ganley [71]. But our proof also shows that the Sylow 2-subgroup of G is cyclic if n is odd.

First, we want to recall the situation that a plane admits a quasiregular group of type 5.1.2 (f). The projective plane is (a, M)- as well as (b, L)-transitive where M is the line joining a and b, and L ($\neq M$) goes through the point a. We will say that planes with this property have the property (E-H) (equivalently, they admit a quasiregular group of type 5.1.2 (f)) to indicate that the plane admits an elation group E and a homology group H. The group G acts sharply transitively on the points which are not on L and M, and it acts sharply transitively on the lines not through a and not through b. These are $n^2 - n$ points and lines. The point set is partitioned into n point classes of size $n - 1$ (the points on the lines through b) and $n - 1$ point classes of size n (the points on the lines through a). Therefore, this incidence structure Π' has the following properties:

(a) Π' consists of $n^2 - n$ points and lines.

(b) Every line has exactly $n - 1$ points.

(c) The point set is divided into n point classes of size $n - 1$ and $n - 1$ point classes of size n such that two points in the same point class are <u>not</u> joined by a line, and if they have no point class in common they are joined by exactly one line.

An incidence structure Π' defined by (a), (b) and (c) has the following additional properties: Through each point, there are exactly $n - 1$ lines (standard counting argument), thus a point p is joined by a line with exactly $(n - 1)(n - 2) = n^2 - n - 1 - (n - 1) - (n - 2)$ points different from p. This shows that there is exactly one point class of size $n - 1$ through p and one point class of size n. Therefore, two point classes of different sizes intersect in exactly one point. This can be used to reconstruct the projective plane Π from Π'. This reconstruction or embedding is unique, the details are left to the reader. Note that the new lines have to be basically the point classes of Π'.

Now let us return to the case of quasiregular groups. The quasiregular group acts sharply transitively as an automorphism group on Π'. The stabilizer of a point on M is a subgroup E of order n which is actually the group of (a, M)-elations. Similarly, the stabilizer of a point on L is a subgroup of order $n - 1$, and H is the group of (b, L)-homologies. The groups E and H are normal subgroups since G is quasiregular. If we identify points of Π' with group elements (after choosing a base point p_0 and a base line L_0 in Π'), then the points on L_0 correspond to a subset R of G ($|R| = n - 1$) with the property that the list of quotients ($\neq 1$) formed by the elements in R contains every group element of G which is not in H or E exactly once. Moreover, no element in H or E occurs as such a quotient. (This follows in exactly the same way as the existence of divisible difference sets from divisible designs with Singer groups. The property that two points are joined by exactly one line translates into the condition that a certain element occurs exactly once as a quotient with elements from a base block.) This property can be reformulated in terms of a group ring equation. The subset R satisfies

$$RR^{(-1)} = n + G - E - H \quad \text{in } \mathbf{Z}[G]. \tag{5.19}$$

Conversely, if a group ring element satisfies (5.19), we can reconstruct the incidence structure Π' (defined by (a), (b) and (c) above) on which G acts sharply transitively. If E (resp. H) is a normal subgroup, then E (resp. H) acts sharply transitively on the points in any point class of size n (resp. $n - 1$). If E (or H) is not normal, then E (or $\overline{H}$) acts only sharply transitively on the point class through the base point p_0: The points $g(p_0)$ (where g runs through E or H) are distinct, and any two points $g(p_0)$ and $h(p_0)$ $(g \neq h)$ are not joined by a line. Otherwise, $g(p_0)$ and $h(p_0)$ are on some line $x(L_0)$ with $g = rx, h = r'x$ $(r, r' \in R)$, and $gh^{-1} = rr'^{-1}$ would be a non-identity element in E (or H) with a quotient representation with elements of R. Note that the other point classes are the conjugates of E and H and hence the conjugate subgroups act sharply transitively on the other point classes. Let us summarize this discussion in the following proposition:

Proposition 5.3.1 *Let Π' be an incidence structure which has properties (a), (b) and (c) as above. The incidence structure Π' admits an automorphism group G acting sharply transitively on points and lines if and only if there is a subset R of G that satisfies (5.19). Here E and H are subgroups of G of order n and $n - 1$. They occur as the stabilizers of the two point classes through some point. If E and H are normal subgroups, then E and H are the stabilizers of all point classes of size n and $n - 1$. In this case, G acts as a quasiregular collineation group of type 5.1.2 (f) in Π where Π is the unique projective plane of order n in which Π' can be embedded.* □

In the special case that $G \cong H \times E$, this proposition is in Ganley [71]. In this case, he called a subset of G that satisfies (5.19) a **direct product difference set** (for short, a d.p. difference set). Then E and H are subgroups of elations and homologies, and G acts as a quasiregular collineation group. In the following, we will always denote the group of elations by E and the group of homologies by H. In the more general context of arbitrary subgroups, E and H are subgroups of *generalized elations* and *generalized homologies*. These are collineations which do not fix an "axis" pointwise but only setwise, see Jungnickel and Vedder [105] and Vedder [170]. An example where H is not normal can be constructed from the desarguesian plane $\mathrm{PG}(2, q)$: Here G is the semi-direct product of E (a normal subgroup of elations) and H (a subgroup of generalized homologies). We take

$$E := \left\{ \begin{pmatrix} 1 & z & 0 \\ 0 & 1 & 0 \\ 0 & 0 & 1 \end{pmatrix} : z \in \mathrm{GF}(q) \right\}, \quad H := \left\langle \begin{pmatrix} 1 & 0 & 0 \\ 0 & x & 0 \\ 0 & 0 & y \end{pmatrix} \right\rangle,$$

where $y \neq 1$ and yx^{-1} is a generator of $GF(q)$. The group G generated by E and H (in $PGL(3, q)$) acts sharply transitively on Π' where the points a and b and the lines M and L are as follows: The point a is $\langle (1\ 0\ 0)^t \rangle$ and the point b is $\langle (0\ 0\ 1)^t \rangle$. The line through a and b is M and the line L through the point a (the "axis" of H) is the line joining a and $\langle (0\ 1\ 0)^t \rangle$. We need yx^{-1} to be a generator of $GF(q)$ in order to make sure that H acts sharply transitively on the lines through the point a which are different from L and M, and therefore G acts sharply transitively on Π'. If $x \neq 1$, the group G is non-abelian (if $x = 1$, then E and H commute). The permutations

$$\{id, (243), (143)\} \subset A_4$$

in the alternating group A_4 give an example where E is the subgroup of order 4, and H is the subgroup generated by (142) in A_4. Another example is

$$R = \{1, u^3 v, u^3 v^2, u^4 v^3\} \quad \text{in} \quad G = \langle u, v : u^5 = v^4 = 1, v^{-1}uv = u^4 \rangle.$$

Now we return to the abelian case. Let H be the multiplicative group of $GF(q)$ and E the additive group of $GF(q)$. Then the set $R = \{(x, x) : x \in H \times E\}$ is a d.p. difference set relative to E and H. We can generalize this example slightly. Let f be an isomorphism of the cyclic group of order $q - 1$ (viewed as the multiplicative group of $GF(q)$) into $GF(q)^*$. We obtain the following example:

Example 5.3.2 The set

$$R = \{(x, f(x)) : x \in GF(q)^*\} \subset GF(q)^* \times (GF(q), +)$$

is a d.p. difference set relative to E and H where E is the <u>additive</u> group and H the <u>multiplicative</u> group of $GF(q)$. For proof, it is enough to check that

$$xy^{-1} = x'y'^{-1} \ \underline{\text{and}} \ f(x) - f(y) = f(x') - f(y') \quad \text{implies} \quad x = x' \text{ and } y = y'$$

(provided $x \neq y$). We write the second equation as

$$f(y)(f(x)f(y)^{-1} - 1) = f(y')(f(x')f(y')^{-1} - 1),$$

hence $f(y)(f(xy^{-1}) - 1) = f(y')(f(x'y'^{-1}) - 1)$. If $xy^{-1} \neq 1$, then $f(xy^{-1}) \neq 1$ and therefore (using $xy^{-1} = x'y'^{-1}$) we get $y = y'$ and $x = x'$.

Let us look more closely at this example. Obviously, any injective function $f : H \to E$ with the property

$$xy^{-1} = x'y'^{-1} \neq 1 \text{ and } f(x) - f(y) = f(x') - f(y') \tag{5.20}$$
$$\text{implies} \quad x = x' \text{ and } y = y'$$

yields a d.p. difference set $R = \{(x, f(x)) : x \in H\}$ in $H \times E$ and, conversely, a d.p. difference set in $H \times E$ gives rise to a function (5.20). This is true since E and H are finite groups, and (5.20) says that every element in $H \times E$ has at most one "difference representation" with elements from R. The function f is said to be **associated with the d.p. difference set**. Our next aim is to characterize the desarguesian planes in the class of planes which can be described by d.p. difference sets. They correspond basically to those functions f which are isomorphisms from the cyclic group of order $n - 1$ into the multiplicative group of the field $GF(n)$. To prove this, we need the concept of multipliers as introduced in Section 1.3: Multipliers are group automorphisms which map the d.p. difference set R onto a translate of R. The d.p. difference sets constructed in the example above admit many multipliers:

Lemma 5.3.3 *Let q be a prime power and f a map $f : H \to GF(q)^*$ with $f(e) = 1$ such that $R = \{(x, f(x)) : x \in H\} \subset H \times (GF(q), +)$ is a d.p. difference set. Then R admits*

$$\varphi_t : \quad H \times (GF(q), +) \quad \to \quad H \times (GF(q), +)$$
$$(x, y) \quad \mapsto \quad (x, ty)$$

as multipliers for each $t \in GF(q)^*$ if and only if H is a cyclic group and f is an isomorphism.

(Here e is the identity element of H. Note that the assumption $f(e) = 1$ is no restriction since we can simply replace R by a suitable translate. The multiplication ty has to be carried out in $GF(q)$.)

Proof. If φ_t is a multiplier, then $\varphi_t(R)$ must be a translate Rh of R. The set R contains no element in H but exactly one element in each of the remaining cosets of H. The same is true for $\varphi_t(R)$ and thus $h \in H$. Furthermore, $\varphi_t(R)$ contains the element (e, t) which shows $h = (f^{-1}(t))^{-1}$. Let y be an arbitrary element of $GF(q)$. We obtain that $(y, tf(y))$ is an element of $Rh = \{(yh, f(y)) : y \in H\}$, hence $(y, tf(y)) = (y, f(yh^{-1}))$ and $f(f^{-1}(t))f(y) = f(yf^{-1}(t))$. Since t is an arbitrary element of $GF(q)^*$, we see that f must be an isomorphism and H is cyclic. For a proof of the other direction, we observe that $\varphi_t(R) = \{(y, tf(y))\} = \{(y, f(sy))\}$ is a translate of R (here $f(s) = t$). $\qquad\square$

Let R_1 and R_2 be two d.p. difference sets with associated functions f_1 and f_2. If f_1 and f_2 are both isomorphisms from H into $GF(q)^*$, then there is obviously a group automorphism $\psi \in \text{Aut}(H \times GF(q)^*)$ such that $R_2 = \psi(R_1)$, i.e., R_1 and R_2 are equivalent. The projective planes corresponding to R_1 and R_2 are isomorphic: There is just one isomorphism class of projective planes whose associated functions are isomorphisms. As one might expect, these are the desarguesian planes. This can be shown as follows: Take

$$E := \{ \begin{pmatrix} 1 & x & 0 \\ 0 & 1 & 0 \\ 0 & 0 & 1 \end{pmatrix} : x \in GF(q)\} \quad \text{and} \quad H := \{ \begin{pmatrix} 1 & 0 & 0 \\ 0 & 1 & 0 \\ 0 & 0 & x \end{pmatrix} x \in GF(q)\},$$

choose the base point $p_0 = \langle (1\ 1\ 1)^t \rangle$ and the base line $L_0 = \langle (1\ 0\ 1)^t, (1\ 1\ 0)^t \rangle$. Then

$$R := \left\{ \begin{pmatrix} 1 & x & 0 \\ 0 & 1 & 0 \\ 0 & 0 & x \end{pmatrix} : x \in \mathrm{GF}(q)^* \right\}$$

is the d.p. difference set and the associated function f is defined by

$$f\left(\begin{pmatrix} 1 & 0 & 0 \\ 0 & 1 & 0 \\ 0 & 0 & x \end{pmatrix} \right) := \begin{pmatrix} 1 & x & 0 \\ 0 & 1 & 0 \\ 0 & 0 & 1 \end{pmatrix}.$$

Note that E is an elementary abelian group which is written "multiplicatively" in $\mathrm{PGL}(3, q)$. We <u>define</u> the field multiplication $*$

$$\begin{pmatrix} 1 & x & 0 \\ 0 & 1 & 0 \\ 0 & 0 & 1 \end{pmatrix} * \begin{pmatrix} 1 & x' & 0 \\ 0 & 1 & 0 \\ 0 & 0 & 1 \end{pmatrix} := \begin{pmatrix} 1 & xx' & 0 \\ 0 & 1 & 0 \\ 0 & 0 & 1 \end{pmatrix}.$$

In this sense, f is an isomorphism.

Theorem 5.3.4 (Pott [149]) *Let R be an abelian d.p. difference set in $H \times \mathrm{GF}(q)$ with associated function f. The corresponding projective plane is desarguesian if and only if H is cyclic and R is equivalent to a d.p. difference set whose associated function is an isomorphism.* □

This characterization of the desarguesian d.p. difference sets will be an important tool in the proof that abelian d.p. difference sets of prime order p describe desarguesian projective planes. We will show that d.p. difference sets of prime order have to admit many multipliers which forces the function f to be an isomorphism. Then Theorem 5.3.4 shows that the planes are desarguesian. In what follows, we think of the d.p. difference set R as an element in $\mathbf{Z}[G]$ (as usual) and consider characters χ from G into $\mathbf{Q}(\zeta_v)$ where v is a primitive complex v-th root of unity, and v is the exponent of the abelian, multiplicatively written group $G \cong H \times E$. The character values satisfy

$$\chi(R)\chi(R^{(-1)}) = \begin{cases} (n-1)^2 & \text{if } \chi = \chi_0 \\ 0 & \text{if } \chi|E = \chi_0, \chi \neq \chi_0 \\ 1 & \text{if } \chi|H = \chi_0, \chi \neq \chi_0 \\ n & \text{otherwise.} \end{cases} \tag{5.21}$$

This follows from equation (5.19) using Lemma 1.2.1. The character values show that $RR^{(-1)}$ is <u>not</u> invertible in $\mathbf{C}[G]$ since there is a character value (equivalently eigenvalue) 0. Therefore, the standard argument used to prove that a multiplier fixes at least one block fails (see Theorem 1.3.8) and our multipliers behave in a somewhat unusual way. If we look at the multipliers in Lemma 5.3.3, it is easy to see that the φ_t's fix no translate of R (if $t \neq 1$): The multiplier φ_t fixes the cosets of H setwise, but it fixes no non-zero element. Since each translate of R meets each coset of H at most once, no translate can be fixed by φ_t.

Theorem 5.3.5 (Pott [149]) *Let R be a d.p. difference set of prime order p in an abelian group G relative to E and H. Then the integers*

$$t_i := p + i(p - 1), \quad i = 1, \ldots, p - 1$$

are numerical multipliers of R.

Proof. In the beginning of the proof, we want to study a situation which is more general than stated in the Theorem. Let p be a prime such that p^e but not p^{e+1} divides n ($e \geq 1$). We define $w' := n(n-1)/p^e$ and let s be a number congruent to a power of p modulo w' with $(s, p) = 1$. Result 1.2.7 (c) (with $w = n(n-1)$) shows that

$$\chi(R^{(s)})\chi(R^{(-1)}) \equiv \chi(R)\chi(R^{(-1)}) \equiv 0 \bmod p^e$$

for characters χ with $\chi|E \neq \chi_0$ and $\chi|H \neq \chi_0$. Since $RE = R^{(s)}E = G$, we can also say $\chi(R) = \chi(R^{(s)}) = 0$ if $\chi|E = \chi_0, \chi \neq \chi_0$. What can we say about non-principal characters χ with $\chi|H = \chi_0$? Let us assume (w.l.o.g.) that R contains no element of H. Then we have $RH = G - H$ and $\chi(RH) = -\chi(H)$, therefore $\chi(R) = -1$ if $\chi|H = \chi_0$ and $\chi \neq \chi_0$. The same is true for $\chi(R^{(s)})$. This shows $\chi(R) = \chi(R^{(s)}) = -1$ if $\chi|H = \chi_0$ but $\chi \neq \chi_0$ and

$$\chi(R^{(s)})\chi(R^{(-1)})) \equiv \chi(R)\chi(R^{(-1)}) \equiv 0 \bmod p^e \tag{5.22}$$

holds for <u>all</u> characters.

Now let us write

$$R^{(s)}R^{(-1)} = G - H + F \tag{5.23}$$

where F is a suitable element in $\mathbf{Z}[G]$. Note that the coefficients of F are greater or equal to -1. From (5.19) and (5.22) we obtain

$$\chi(F) \equiv \chi(n - E) \equiv 0 \bmod p^e \tag{5.24}$$

for all characters.

Next, we compute $R^{(s)}R^{(-1)}R^{(-s)}R$ in the two ways $(R^{(s)}R^{(-1)})(R^{(-s)}R)$ and $(R^{(s)}R^{(-s)})(RR^{(-1)})$ and obtain

$$(G - H + F)(G - H + F^{(-1)}) = (G - H + (n - E))(G - H + (n - E)).$$

Note that one can easily compute $FG = (n - E)G = 0$ and $FH = (n - E)H$ using (5.23) since $RG = (n-1)G$ and $RH = G - H$. This shows

$$FF^{(-1)} = (n - E)^2 = n^2 - nE \tag{5.25}$$

where $F \in \mathbf{Z}[G]$ and $\chi(F) \equiv 0 \bmod p^e$ for all characters of G, see (5.24). If we can prove that the only possible solution for F in (5.25) is $F = (n - E)k$ for some $k \in G$, we can prove that s is a multiplier: We compute

$$(R^{(s)}R^{(-1)})R = (G - H + nk - Ek)R = (R^{(s)}(R^{(-1)}R)) = R^{(s)}(n + G - H - E)$$

which shows $R^{(s)} = Rk$ (since $RE = R^{(s)})E = G$) and thus s is a multiplier. I think that it is worth to mention this explicitly since there might be techniques not yet known to me to prove that (5.25) has only solutions of the form $F = (n - E)k$. At present, I can prove this only if $n = p$, hence if E is a cyclic group of prime order p. In this case, the integers s which are relatively prime to p and congruent p modulo $p - 1$ are precisely the t_i's of the form $t_i = p + i(p - 1)$. In order to prove $F = (p - E)k$, we have to involve Lemma 1.2.14. Let $F = \sum f_g g$, then Lemma 1.2.14 and (5.24) tell us that the f_g's are congruent modulo p on cosets of E. The coefficient of the identity in $FF^{(-1)}$ is

$$\sum_{g \in G} f_g^2 = p^2 - p. \tag{5.26}$$

Since $FF^{(-1)} = p^2 - pE$ is <u>not</u> constant on cosets of E, there must be at least one coset Ek such that $f_g \neq f_h$ for suitable $g, h \in Ek$. Let y be the integer in the interval $[0, p - 1)$ with $f_g \equiv y \bmod p$ for all $g \in Ek$. Let us first assume that $y \neq p - 1$. Then there is at least one coefficient $f_g \geq p$ (note $f_g \geq -1$) contradicting (5.26). If $y = p - 1$, we have $f_g = p - 1$ for some $g \in Ek$ and $f_h = -1$ for $h \in Ek$, $h \neq g$. Using $\sum_{g \in Ek} f_g^2$ and (5.26), we see that $f_g = 0$ for $g \notin Ek$, which proves our theorem. $\qquad \square$

Corollary 5.3.6 *A projective plane of prime order p with a quasiregular collineation group of type 5.1.2 (f) must be a desarguesian plane.*

Proof. We just have to combine Lemma 5.3.3 and Theorems 5.3.4 and 5.3.5: The multipliers φ_t in Lemma 5.3.3 are precisely the multipliers constructed in Theorem 5.3.5. Then Theorem 5.3.4 shows that the plane has to be desarguesian. $\qquad \square$

Note that our theorem looks quite analogous to Result 5.4.8. However, the proofs are very different. In the case of quasiregular groups of type (b), we cannot use multipliers to prove something since there is no multiplier theorem that tells us which numbers are multipliers.

Our next aim is to give strong numerical restrictions on the orders n of abelian d.p. difference sets. I think that we are close to prove the prime power conjecture in this situation. We begin with the n even case. We need tools similar to those in the previous section. Assume that a projective plane Π is (a, M)- as well as (b, L)-transitive where M is the line joining a and b and L is a line through the point a which is different from M. Let G be the quasiregular collineation group generated by these elations and homologies. If G is abelian, then Π admits a polarity. In order to define this polarity, we choose a (base) point p which is not on L or M, and a (base) line B which does not contain a

or b, thus B corresponds to a d.p. difference set. The map

$$
\begin{array}{ccccc}
g(p) & \mapsto & g^{-1}(B), & \qquad g(B) & \mapsto & g^{-1}(p) \\
g(B \cap M) & \mapsto & g^{-1}(pa), & \qquad g(pa) & \mapsto & g^{-1}(B \cap N) \\
g(B \cap L) & \mapsto & g^{-1}(pb), & \qquad g(pb) & \mapsto & g^{-1}(B \cap L) \\
a & \mapsto & M, & \qquad M & \mapsto & a \\
b & \mapsto & L, & \qquad L & \mapsto & b
\end{array}
\tag{5.27}
$$

(where $g \in G$) is a polarity. This is already contained in Ganley [71] and Ganley
and Spence [73]. It is straightforward to check that this map is a polarity
(Proposition 5.2.2). Another look at this situation is the following: We consider
the residual incidence structure Π' (as introduced earlier) and denote by R the
d.p. difference set corresponding to base line B and base point p. Then the
interchange

$$ g \leftrightarrow Rg^{(-1)} $$

of points and lines is a polarity if G is abelian. Note that $g \in Rh$ if and only if
$h^{-1} \in Rg^{-1}$. We emphasize that this argument works only if G is abelian. (Here
a polarity of Π' is of course an isomorphism of order two between Π' and its dual
$(\Pi')^t$.) Therefore, the dual of Π' is isomorphic to Π'. Since the extensions of
Π' (say Π) and $(\Pi')^t$ are unique, the extension of Π' must be isomorphic to
the extension of $(\Pi')^t$. The isomorphism of order two between Π' and its dual
extends to an isomorphism (hence polarity) between Π and Π^t. The images
of $g(B \cap M)$, $g(B \cap L)$, a and b have to be what they are in (5.27). In the
following theorem, we will use these polarities in the same way as in Theorem
5.2.5. We will count the number of absolute points of the polarities using the
same averaging argument. As already mentioned, part of this theorem (the case
if n is even) is contained in Ganley [71].

Theorem 5.3.7 (Pott [149]) *Let Π be a projective plane of order n admitting
an abelian quasiregular collineation group $G \cong H \times E$ of type 5.1.2 (f). If n is
even, then n is a power of two and the subgroup E of order n is an elementary
abelian 2-group. If n is odd, then the Sylow 2-subgroup of G is cyclic.*

Proof. We will use the notation as in the definition of the polarities (5.27).
We obtain $n(n-1)$ such polarities by choosing different p's but the same base
line B. Since each polarity has at least $n+1$ absolute points (Result 5.2.3),
the total number of absolute points of these polarities is consequently at least
$n(n-1)(n+1)$ with equality if and only if each polarity has exactly $n+1$ absolute
points.

In all the $n(n-1)$ cases, the point a is absolute and no other point on the line
M is absolute (note that G acts transitively on M). The point $g(p)$ is absolute
if and only if $g(p) \in g^{-1}(B)$, equivalently $g^2(p) \in B$. In order to determine the
total number of absolute points $g(p)$ of the $n(n-1)$ polarities, we have to count
the number of triples (q, p, g) such that $g^2(p) = q$ is on the line B. Since the
point p is determined by g (given q), this number is $n(n-1)(n-1)$. Finally, we
must count the number of absolute points on L. The point $g(B \cap L)$ is absolute

if and only if $g^2(B \cap L)$ is on the line through p and b. Now the group G does not act regularly on L but the group E does. So we count the number of pairs (p, g) such that $g^2(B \cap L) = pb \cap L$, $g \in E$. Since g determines $pb \cap L$, there are $n - 1$ choices for p given g. So we get $n(n - 1)$ absolute points on L. All together, we have

$$n(n - 1) + n(n - 1) + n(n - 1)(n - 1) = n(n - 1)(n + 1)$$

absolute points, thus each polarity has exactly $n + 1$ absolute points.

If n is even, the $n + 1$ absolute points form a line (Result 5.2.3). Choose p on the line joining $B \cap L$ and b. Then the points a and $B \cap L$ are absolute points and henceforth all the points on L are absolute. This means $g^2(B \cap L) = B \cap L$ for all $g \in E$, thus E must be an elementary abelian 2-group which proves the first part of the theorem.

If n is odd, the set of absolute points is an oval. Assume that the Sylow 2-subgroup of G is not cyclic. Then there are at least two elements of order 2 in G, say t and s. Let us choose the point p on B. The corresponding polarity has at least three absolute points on B, namely, the points p, $t(p)$ and $s(p)$, a contradiction. □

For the case that n is even this is quite a strong result since it proves the prime power conjecture. As far as I see, the proof cannot be generalized to the n odd case. In that case, we can use projection arguments similar to those in Theorem 5.2.13. However, in the present situation we obtain much stronger results. We need the result by Camion and Mann [40] on antisymmetric difference sets (Result 5.2.14) and the following result on "partial difference sets" due to Ma [123].

Result 5.3.8 (Ma [123]) *Let G be an abelian group of order n where n is a non-square which is congruent 1 modulo 4. Let D be a subset of G that does not contain the identity element of G. If the corresponding group ring element D in $\mathbf{Z}[G]$ satisfies*

$$D = D^{(-1)}$$

and

$$D^2 = \frac{n-1}{4} + \frac{n-1}{4}G - D \quad \text{in } \mathbf{Z}[G],$$

then n is a power of some prime p with $p \equiv 1 \mod 4$.

Theorem 5.3.9 (Pott [149]) *Let R be an abelian d.p. difference set of non-square order n in G relative to E and H where $|E| = n$ and $|H| = n - 1$, and where n is odd. Then n is a prime power.*

Proof. We begin with the equation

$$RR^{(-1)} = n + G - E - H$$

and we assume w.l.o.g. that $R \cap H = \emptyset$. We factor out a subgroup of order $(n-1)/2$ in H and denote the canonical epimorphism by ρ. We obtain

$$\rho(R)\rho(R^{(-1)}) = n + \frac{n-1}{2}\rho(G) - \rho(E) - \frac{n-1}{2}\rho(H) \qquad (5.28)$$

with $\rho(E) \cong E$, $\rho(H) \cong \{1, i\}$ (the cyclic group of order 2), and $\rho(G) \cong \rho(E) \times \{1, i\}$. Let us say that $\rho(E) = E$. We can write $\rho(R)$ as $A + Bi$ for $A, B \in \mathbf{Z}[E]$. Equation (5.28) shows

$$AA^{(-1)} + BB^{(-1)} = \frac{n+1}{2} + \frac{n-3}{2}E. \qquad (5.29)$$

Since the projection from G onto $\rho(G)$ maps maps R onto $E \setminus \{1\}$ (note $R \cap H = \emptyset$), we see that A and B are group ring elements with 0/1-coefficients and $A + B + 1 = E$. Let us define $a := |A|$ and $b := |B|$. We have

$$a + b = n - 1$$

and

$$a^2 + b^2 = (n+1)/2 + n(n-3)/2 = (n-1)^2/2$$

which implies

$$a = b = (n-1)/2.$$

We replace B by $E - A - 1$ in (5.29) and get

$$2AA^{(-1)} + nE + 1 - 2E - (n-1)E + A + A^{(-1)} = \frac{n+1}{2} + \frac{n-3}{2}E,$$

therefore

$$2AA^{(-1)} + A + A^{(-1)} = \frac{n-1}{2} + \frac{n-1}{2}E. \qquad (5.30)$$

Case (a) $n \equiv 3 \bmod 4$. We consider (5.30) modulo 2 and obtain (note $1 \notin A$)

$$A + A^{(-1)} = E + 1 \quad \text{in GF}(2)[E]$$

which shows $A \cap A^{(-1)} = \emptyset$ and $A^{(-1)} = B$. Using (5.29), we see

$$AA^{(-1)} = \frac{n+1}{4} + \frac{n-3}{4}E \quad \text{in } \mathbf{Z}[E]$$

and A must be an antisymmetric $(n, (n-1)/2, (n-3)/4)$-difference set. Result 5.2.14 shows that n must be a prime power.

Case (b) $n \equiv 1 \bmod 4$. Again, we consider (5.30) modulo 2 to get $A + A^{(-1)} = 0$, hence $A = A^{(-1)}$. Then (5.30) reduces to

$$A^2 = \frac{n-1}{4} + \frac{n-1}{4}E - A \quad \text{in } \mathbf{Z}[E]. \qquad (5.31)$$

Result 5.3.8 shows that n is a prime power if n is not a square. This finishes the proof. $\qquad \square$

It is probably true that the assumption "n is not a square" in Result 5.3.8 is not necessary. Sometimes it is conjectured that n has to be a prime power. I think that it might be possible to solve the case that n is a square in Theorem 5.3.9, no matter whether Result 5.3.8 remains true for squares or not. There might be an induction argument around. If n is a square, say $n = m^2$, then an (abelian) quasiregular group of order $m^2(m^2 - 1)$ has a subgroup of order $m(m - 1)$. If it is possible to show that the orbit of some point under this subgroup "is" a Baer subplane, then induction is possible. In the classical desarguesian case it is true that a subgroup of order $m(m - 1)$ induces a subplane.

A final remark: Examples of antisymmetric difference sets and the partial difference sets in Result 5.3.8 are provided by the sets of squares in $\mathrm{GF}(q)$. If we start with the classical d.p. difference set $\{(x, x) : x \in \mathrm{GF}(q)^*\} \subset \mathrm{GF}(q)^* \times (\mathrm{GF}(q), +)$, the projection procedure in Theorem 5.3.9 yields exactly the squares in $\mathrm{GF}(q)$. To see this, note that the unique subgroup of order $(q-1)/2$ in $\mathrm{GF}(q)^*$ is the set of squares.

Up to now, the only <u>known</u> examples of antisymmetric difference sets are the (non-zero) squares or non-squares. On the other hand, Davis [52] and Leung and Ma [118] recently constructed new examples of partial difference sets of "Paley type" (i.e., solutions of (5.31)) with $A = A^{(-1)}$: Examples are known to exist also in $\mathbf{Z}_{q^2} \times \mathbf{Z}_{q^2}$ whenever q is an odd prime power.

5.4 Planar functions

In this section, we will summarize some interesting results about $(n, n, n, 1)$-difference sets. Again, we can use our general theory about divisible difference sets to get informations about these semiregular difference sets with $\lambda = 1$, see, for instance, Proposition 2.4.3. In Chapter 4, we have already obtained quite a few results about the case $n = p^a$. We will begin the discussion of the case "$\lambda = 1$" with one of the most interesting results about $(n, n, n, 1)$-difference sets.

Result 5.4.1 (Ganley [70]; Jungnickel [97]) *Assume that an abelian relative $(n, n, n, 1)$-difference set exists in G relative to N. If n is even, then n is a power of 2, $G \cong \mathbf{Z}_4 \times \ldots \times \mathbf{Z}_4$, and $N \cong \mathbf{Z}_2 \times \ldots \times \mathbf{Z}_2$.*

This result shows that the prime power conjecture is true for a large class of planes. In the odd order case, we have the Mann test and the following result which cannot be proved using Theorem 2.4.1 and its variations.

Result 5.4.2 (Hiramine [85]) *There is no splitting abelian relative $(n, n, n, 1)$-difference set if $n = 3p$ for some prime p.*

Note that this result rules out the existence of splitting examples with $n = 39$ and $n = 93$ which has been open in Proposition 2.4.3. The smallest open case today seems to be 55:

Problem 11 Is it true that no splitting relative $(55, 55, 55, 1)$-difference set exists?

From now on, we can restrict ourselves to the odd case (at least, if G is abelian). Let us begin with two immediate corollaries of former results:

Corollary 5.4.3 *Let R be a splitting $(n, n, n, 1)$-difference set in $\mathbf{Z}_n \times \mathbf{Z}_n$. Then n is a square-free number, i.e., n is the product of distinct primes.*

Proof. We assume that p^2 divides n and apply Theorem 4.1.3. $\square$

Corollary 5.4.4 (Nakagawa [138]) *Let R be a splitting abelian $(p^a, p^a, p^a, 1)$-difference set in $H \times N$ relative to N where p is an odd prime. Then the exponent of N satisfies*

$$exp(N) \leq \begin{cases} p^{(n+1)/2} & \text{if } n \text{ is odd} \\ p^{n/2} & \text{if } n \text{ is even.} \end{cases}$$

Proof. This follows from Theorems 4.1.4 and 4.1.5. $\square$

The case of splitting semiregular relative difference sets is of particular interest in view of the connection to group invariant generalized Hadamard matrices, see Proposition 2.2.7. In the special case $\lambda = 1$, these matrices have been called "ideal" matrices, see Kumar [112]. There is another reason why the splitting case with $\lambda = 1$ has attracted the interest of geometers. If $G \cong H \times N$ contains a splitting $(n, n, n, 1)$-difference set R relative to N, then there is a mapping $f : H \rightarrow N$ where $f(h)$ is the (uniquely determined) element in N such that $(h, f(h)) \in R$ (note that R meets each coset of N exactly once). This function has the property that the mapping defined by

$$f_g : \begin{array}{ccc} N & \rightarrow & N \\ h & \mapsto & f(gh)f(h^{-1}) \end{array} \tag{5.32}$$

is a bijection if $g \neq 1$. This follows quite easily from the fact that R is a relative difference set: The quotients $x = f(gh)f(h)^{-1}$ yield the elements (g, x) in G that can be formed as quotients with elements from R. Conversely, a mapping $f : H \rightarrow N$ with the property (5.32) yields a splitting $(n, n, n, 1)$-difference set in $H \times N$ relative to N. A function with this property is called a **planar function**.

Example 5.4.5 We take the $(3, 3, 3, 1)$-difference set

$$(0, 0), \ (1, 1), \ (2, 1)$$

in $\mathbf{Z}_3 \times \mathbf{Z}_3$. The associated planar function is the mapping $f : x \mapsto x^2$ (calculations are carried out in GF(3)). In general, the mapping

$$f : \begin{array}{ccc} \text{GF}(q) & \rightarrow & \text{GF}(q) \\ x & \mapsto & x^2 \end{array}$$

is always a planar function between the <u>additive</u> groups of GF(q) where q is odd (although the definition of the function uses the multiplicative structure of the field).

We can generalize this example:

Proposition 5.4.6 (Dembowski, Ostrom [60]) *Let $q = p^e$ be a power of the odd prime p. Let $a_{i,j}$ be elements in GF(q) such that*

$$\sum_{i,j=0}^{e-1} a_{i,j}(x^{p^i} y^{p^j} + x^{p^j} y^{p^i}) = 0 \quad \text{if and only if} \quad x = 0 \text{ and } y = 0.$$

Then the mapping

$$
\begin{aligned}
f: \quad GF(q) \quad &\to \quad GF(q) \\
x \quad &\mapsto \quad \sum_{i,j=0}^{e-1} a_{i,j} x^{p^i + p^j}
\end{aligned}
$$

is a planar function from the additive group of GF(q) onto itself.

Proof. The proof is a straightforward calculation. □

Every <u>known</u> planar function is equivalent to one described in this Proposition (where we say that planar functions are equivalent if the corresponding difference sets are equivalent). The planes corresponding to quadratic functions have to be desarguesian:

Proposition 5.4.7 *Let q be an odd prime power. Then the projective plane corresponding to the planar function*

$$
\begin{aligned}
f: \quad GF(q) \quad &\to \quad GF(q) \\
x \quad &\mapsto \quad x^2
\end{aligned}
$$

is desarguesian.

Proof. We take the matrix group

$$
\left\{
\begin{pmatrix}
1 & a & -b \\
0 & 1 & 2a \\
0 & 0 & 1
\end{pmatrix}
: a, b \in GF(q)
\right\}.
$$

It is easy to see that this group acts on the desarguesian plane of order q as a quasiregular group of type (b) in Result 5.1.2: The subgroup N will be the group satisfying $a = 0$, the complement H will be the subgroup with $b = -a^2$. The subgroups H and N are labelled with elements from the field GF(q), hence we can think of H and N as "fields". If we take the base point $\langle(1\ 1\ 1)^t\rangle$ and the base line consisting of vectors $\langle(x\ y\ z)^t\rangle$ then the corresponding planar function is "squaring". □

It is not true that all the planar functions in Proposition 5.4.6 describe desarguesian planes. Every projective plane which can be coordinatized by a so called *semifield* can be also described by an $(n, n, n, 1)$-difference set in G. (We refer the reader to the books by Dembowski [59], Hughes and Piper [91], and Pickert [139] for more about semifields and related topics.) Here the group G is abelian if and only if the multiplication in the semifield is commutative, see Hughes [88] and Jungnickel [95]. The planes coordinatized by semifields are also translation planes. The translation planes in the class of all planes which admit planar functions (or $(n, n, n, 1)$-difference sets) have been characterized by Dembowski and Ostrom in terms of planar functions, i.e., there is a condition on the planar function which has to be satified in order that the plane is a translation plane. The functions defined in Proposition 5.4.6 satisfy this property, hence the planes are translation planes (but not necessarily desarguesian). However, all semifields of order p or p^2 are actually fields (Theorem 5.3.10 in Dembowski [59]), i.e., the construction of Hughes using semifields cannot yield non-desarguesian examples. But it might be possible that there are other constructions! In the prime case, however, this is impossible.

Result 5.4.8 (Gluck [75]; Hiramine [83]; Rónyai, Szönyi [154]) *Let R be a relative $(p, p, p, 1)$-difference set in G. Then $G \cong \mathbf{Z}_p \times \mathbf{Z}_p$ and the corresponding projective plane is desarguesian.*

Problem 12 Is it true that an (abelian) relative $(p^2, p^2, p^2, 1)$-difference set corresponds to a desarguesian plane?

In a first step to answer this question, we are able to prove that the group containing the difference set has to be elementary abelian (provided that the group is already abelian). The exponent bounds developed so far show that there are only four possible candidates for abelian groups which contain relative $(p^2, p^2, p^2, 1)$-difference sets.

(1) $G \cong \mathrm{EA}(p^4)$,

(2) $G \cong \mathbf{Z}_{p^2} \times \mathbf{Z}_{p^2}$, $N \cong \mathbf{Z}_p \times \mathbf{Z}_p$,

(3) $G \cong \mathbf{Z}_{p^2} \times \mathbf{Z}_p \times \mathbf{Z}_p$, $N \cong \mathbf{Z}_p \times \mathbf{Z}_p$, $G/N \cong \mathbf{Z}_{p^2}$,

(4) $G \cong \mathbf{Z}_{p^2} \times \mathbf{Z}_p \times \mathbf{Z}_p$, $N \cong \mathbf{Z}_p \times \mathbf{Z}_p$, $G/N \cong \mathbf{Z}_p \times \mathbf{Z}_p$.

The case of a cyclic group or a group with exponent p^3 is impossible because of our exponent bound in Theorem 4.1.5. This theorem also shows that N has to be elementary abelian.

To prove that only the first case can occur, we need the following lemma:

Lemma 5.4.9 *Let R be an abelian relative (p^2, p, p^2, p)-difference set in G relative to N and let U be a subgroup of G of order p with $|U \cap N| = 1$ and*

$G/U \cong \mathbf{Z}_{p^2}$. We denote the ·canonical epimorphism from G onto G/U by ρ. Then there is an element $\rho(g)$ in G/U such that

$$\rho(R) = \rho(g)(\rho(G) - \rho(N) + p).$$

Proof. The characters of G/U are the characters of G which are principal on U. We have $\chi(R) \equiv 0 \bmod p$ for all characters $\chi \in G^*$, hence for all characters of $\rho(G)$ (note that p is self-conjugate modulo $|G|$.) The element $\rho(R)$ has integer coefficients between 0 and p and satisfies

$$\rho(R)\rho(R)^{(-1)} = p^2 + p^2\rho(G) - p\rho(N). \qquad (5.33)$$

We write $\rho(R) = \sum r_g g$ and get

$$\sum r_g = p^2,$$

$$\sum (r_g)^2 = 2p^2 - p.$$

Let μ be the canonical epimorphism from $\rho(G)$ onto $\rho(G)/\rho(N)$. Then $\mu(\rho(R)) = p\mu(\rho(G))$, i.e.

$$\sum_{g \in z\rho(N)} r_g = p \quad \text{for all } z \in \rho(G).$$

Since $\rho(N)$ is the unique subgroup of $\rho(G)$ of order p, the coefficients of $\rho(R)$ are constant modulo p on cosets of $\rho(N)$; this follows from Lemma 1.2.14. The only possibilities for the r_g's with $g \in z\rho(N)$ are

$$r_g = 1 \quad \text{for all } g \in z\rho(N) \qquad (5.34)$$

or

$$r_g = \begin{cases} p & \text{for} \quad g = k \\ 0 & \text{for} \quad g \neq k \end{cases} \qquad (5.35)$$

for some $k \in z\rho(N)$. If x denotes the number of cosets with (5.34) and y the number of cosets with (5.35), we get $x + y = p$ and $px + p^2y = 2p^2 - p$ (coefficient of the identity in (5.33)) which shows $y = 1$ and $x = p - 1$ and proves the lemma. $\square$

Theorem 5.4.10 (Ma, Pott [127]) *Let R be a an abelian relative $(p^2, p^2, p^2, 1)$-difference set in G where p is an odd prime. Then G has to be the elementary abelian group.*

Proof. We will exclude the cases (2), (3), and (4) above.
Case (2). Let U be a subgroup of G of order p^2 with $U \cong \mathbf{Z}_{p^2}$ and $G/U \cong \mathbf{Z}_{p^2}$. The projection epimorphism $\rho : G \to G/U$ can be decomposed

$$G \xrightarrow{\mu_1} G_1 \cong G/U_1 \xrightarrow{\mu_2} G_2 \cong G_1/U_2$$

where U_1 is a subgroup of order p in N and U_2 is a subgroup of order p in G_1 with $|\mu_1(N) \cap U_2| = 1$. The image of the relative difference set R under μ_1 is a relative (p^2, p, p^2, p)-difference set in $H \times U_2$ where $H \cong \mathbf{Z}_{p^2}$, $U_2 \cong \mathbf{Z}_p$ and the forbidden subgroup is a subgroup of H. Lemma 5.4.9 shows that

$$\mu_2(\mu_1(R)) = \rho(R) = \rho(g)(p + \rho(G) - \rho(N)) \tag{5.36}$$

where we may assume (w.l.o.g.) that $g = 1$.

We define $E := R \cap U$. Note that E is the pre-image of p in (5.36) under $\rho|R$, i.e., E is a p-subset of U. Moreover, E satisfies

$$EE^{(-1)} = p + \sum_{g \in U \setminus N} r_g g \quad \text{in } \mathbf{Z}[U]$$

with $r_g \in \{0, 1\}$ since $|E| = p$ and E is part of the relative difference set R. Since $|U| = p^2$ and $|U \cap N| = p$, we conclude that E is a relative $(p, p, p, 1)$-difference set in the cyclic group U which is impossible if p is odd, see Theorem 4.1.1.

Case (3) and (4). Let U be a subgroup of G such that $U \cong \mathbf{Z}_p$ and $|U \cap N| = 1$. Note that such a subgroup U always exist. Let $\rho : G \to G/U \ (=: G_1)$ be the canonical epimorphism. Then

$$\rho(R)\rho(R)^{(-1)} = p^2 + pG_1 - N_1$$

where $G_1 \cong \mathbf{Z}_{p^2} \times \mathbf{Z}_p$ and $N_1 := \rho(N) \cong \mathbf{Z}_p \times \mathbf{Z}_p$. Now let K be a subgroup of N_1 (and hence a subgroup of N) with $G_1/K \cong \mathbf{Z}_{p^2}$, and let μ be the canonical epimorphism $G_1 \to G_1/K$. Then there exists some $h \in G$ such that

$$\mu(\rho(R)) = \mu(\rho(h))(G_1/K - N_1/K + p) \tag{5.37}$$

since $\mu(\rho(R))$ is the image (in a cyclic group of order p^2) of a relative (p^2, p, p^2, p)-difference set in $G/K \cong \mathbf{Z}_{p^2} \times \mathbf{Z}_p$ (Lemma 5.4.9). We may assume (w.l.o.g.) that $h = 1$.

Let $H = \rho^{-1}(K) \cong \mathbf{Z}_p \times \mathbf{Z}_p$ and $E := R \cap H$. As in case (2), the set E is a relative $(p, p, p, 1)$-difference set in H with respect to $H \cap N \ (= K)$. Note that $H = U \times K$ and we will not distinguish between K and $\rho(K)$. We obtain

$$\rho(E)\rho(E)^{(-1)} = p + (p - 1)K. \tag{5.38}$$

This equation shows that there is at least one coefficient in $\rho(E)$ greater than 1. Let $\rho(R) = \sum_{g \in G_1} r_g g$. By (5.36), we get

$$r_g = \begin{cases} 0 & \text{if } g \in N_1 \setminus K \\ 0 \text{ or } 1 & \text{if } g \in G_1 \setminus N_1 \\ > 1 & \text{for some } g \in K. \end{cases} \tag{5.39}$$

Now we repeat the same argument with another subgroup $L \ (\neq K)$ of N_1, $L \cong \mathbf{Z}_p$, and $G_1/L \cong \mathbf{Z}_{p^2}$. Here we cannot assume $h = 1$ in (5.37), again. But we can conclude that some r_g has to be greater than 0 for some $g \in Lh \ (h \in \rho(G))$ and

$$r_g = \begin{cases} 0 & \text{if } g \in N_1 h \setminus Lh \\ 0 \text{ or } 1 & \text{if } g \in G_1 \setminus N_1 h \\ > 1 & \text{for some } g \in Lh. \end{cases} \tag{5.40}$$

If $N_1 h \neq N_1$, then (5.39) and (5.40) cannot be both satisfied, hence $N_1 h = N_1$ and the two sets Lh and K intersect in some element h'. Therefore, the coefficient $r_{h'}$ is the only coefficient of $\rho(E)$ which is different from 0, hence $r_{h'} = p$ contradicting (5.38). □

There are a few more results about the "structure" of a group which contains an $(n, n, n, 1)$-difference set. The proofs given here are, in my opinion, easier and shorter than the original ones:

Theorem 5.4.11 (Hiramine [84]) *Let R be an abelian $(n, n, n, 1)$-difference set in G relative to N. Then the following holds:*

(a) If the rank of G is 2, then $G \cong \mathbf{Z}_n \times \mathbf{Z}_n$.

(b) Let p be an odd prime, and let S be a Sylow p-subgroup of G. If G/N has a cyclic Sylow p-subgroup, then $S \cap N$ is a direct factor of S.

Proof. The first statement is immediate from Theorem 4.1.1.

Now let p^i be the largest power of p dividing n. We choose an element $g \in S$ such that the coset gN has order p^i in G/N. This is possible since G/N has a cyclic Sylow p-subgroup. We have $g^{p^i} \in N$. If $g^{p^i} \neq 1$, there would be an element outside N (namely g) of order not dividing n contradicting Theorem 4.1.1 (note that we may assume that n is odd in view of Result 5.4.1). This shows that $g^{p^i} = 1$, and thus g generates a subgroup of order p^i which is a complement of $S \cap N$ in S. □

Chapter 6

Codes and sequences

In this final chapter, we will investigate the connection between difference sets (in particular, cyclic difference sets), periodic sequences, and codes. In Section 6.1, we will consider the question whether "perfect" sequences can exist. Since this is (with only a few exceptions) presumably never the case, we will consider so called "almost perfect sequences" in Section 6.2. Finally, we will determine the dimensions of certain codes defined via difference sets, and we will solve a question posed by Assmus and Key on ovals in the code of a desarguesian plane.

6.1 Perfect sequences

Important parameters of periodic sequences are the periodic and aperiodic auto-correlation coefficients. Let $a = (a_i)$ be a periodic sequence over K with period v. Then the **aperiodic autocorrelation coefficients** are

$$\alpha_s(a) := \sum_{i=0}^{v-1-s} a_i a_{i+s},\tag{6.1}$$

the **periodic autocorrelation coefficients** are

$$\gamma_s(a) := \sum_{i=0}^{v-1} a_i a_{i+s}\tag{6.2}$$

where we reduce the subscripts modulo v. Obviously, we have

$$\gamma_s(a) = \alpha_s(a) + \alpha_{v-s}(a)$$

for $0 \le s \le v - 1$. The case $s = 0$ or, more generally, s is a multiple of v, is a special case, and we call these autocorrelation coefficients the trivial ones. We say that a sequence is **perfect** if all its non-trivial periodic autocorrelation coefficients are 0. A **Barker sequence** of length v is a periodic sequence of period v with entries ± 1 such that all non-trivial aperiodic autocorrelation coefficients are 0 (considered in $\mathbf{Z}$). If all the non-trivial periodic autocorrelation coefficients γ_s of a Barker sequence are equal to some constant γ, the Barker sequence is called a **periodic Barker sequence**.

Example 6.1.1 The integer sequence $(-1, 1, 1, 1, \ldots)$ with period 4 is a perfect sequence. Another example with period 7 is $(1, 0, 0, -1, 0, -1, -1, \ldots)$. The following "sequences" are periodic Barker sequences:

- $(-1, 1, 1, 1)$, period 4, $\gamma = 0$;

- $(1, 1, -1)$, period 3, $\gamma = -1$;

- $(1, 1, 1, -1, -1, 1, -1)$, period 7, $\gamma = -1$;

- $(1, 1, 1, -1, -1, -1, 1, -1, -1, 1, -1)$, period 11, $\gamma = -1$;

- $(1, 1, 1, 1, 1, -1, -1, 1, 1, -1, 1, -1, 1)$, period 13, $\gamma = 1$.

It is by no means an incidence that the Barker sequences in the example above are periodic. One can prove:

Result 6.1.2 (Turyn, Storer [169]) *A Barker sequence is periodic.*

Turyn and Storer could also prove the following interesting result:

Result 6.1.3 (Turyn, Storer [169]) *There are no Barker sequences of odd length v with $v > 13$.*

One can also show that the list of examples in 6.1.1 is complete in the odd order case (up to "equivalence"). Therefore, we just have to look at the even length case. Before we do this, we will investigate the connection between cyclic difference sets and the autocorrelation coefficients.

Proposition 6.1.4 *Let D be a cyclic $(v, k, \lambda; n)$-difference set (i.e., D is a set of residues modulo v). We define a sequence $a := (a_i)$ with period v via*

$$a_i = \begin{cases} 1 & \text{if } i \in D \\ -1 & \text{if } i \notin D \end{cases}$$

Then the periodic autocorrelation coefficients with $0 \leq s \leq v - 1$ are

$$\gamma_s(a) = \begin{cases} v & \text{if } s = 0 \\ v - 4n & \text{if } s \neq 0 \end{cases} \tag{6.3}$$

Conversely, if a sequence with period v and coefficients ± 1 has constant (non-trivial) periodic autocorrelation coefficients γ, then the set

$$D := \{i : a_i = 1\}$$

considered modulo v is a cyclic $(v, k, \lambda; n)$-difference set. The value k is the number of residues with $a_i = 1$ and $\lambda = (\gamma + 4k - v)/4$.

Proof. To prove the first part of the theorem, we consider the element D in $\mathbf{Z}[G]$. The group ring element

$$A := \sum_{i=0}^{v-1} a_i i$$

is simply $2D - G$. The periodic autocorrelation coefficients are nothing else than the coefficients of $AA^{(-1)}$ in $\mathbf{Z}[G]$ (here we write the group multiplicatively!). We compute

$$AA^{(-1)} = (2D - G)(2D^{(-1)} - G) = 4(n + \lambda G) - 4kG + vG = 4n + (v - 4n)G$$

which proves the validity of (6.3). Conversely, we have

$$AA^{(-1)} = (v - \gamma) + \gamma G.$$

The set D is $(A + G)/2$ which shows

$$
\begin{aligned}
4DD^{(-1)} &= (A + G)(A^{(-1)} + G) \\
&= (v - \gamma) + \gamma G + vG + 2(k - (v - k))G \\
&= (v - \gamma) + (\gamma + 4k - v)G.
\end{aligned}
$$

Therefore, D is a difference set with the desired parameters.   □

This proposition shows that cyclic difference sets and sequences with constant autocorrelation coefficients are basically the same objects! Regarding sequences where γ is small (which are needed in engineering applications), we can say the following:

Corollary 6.1.5 *The existence of a perfect sequence with entries ± 1 and period v is equivalent to the existence of a cyclic Hadamard difference set in $\mathbf{Z}_v$. The existence of an even length Barker sequence implies the existence of a cyclic Hadamard difference set. Sequences with constant autocorrelation $\gamma = -1$ are equivalent to cyclic $(4n - 1, 2n - 1, n - 1; n)$-difference sets (Paley-Hadamard difference sets) . If $\gamma = 1$, then the corresponding cyclic difference sets have parameters $(2t(t + 1) + 1, t^2, t(t - 1)/2; t(t + 1)/2)$.*

Proof. The only statement which is not immediate is the assertion about the parameters of the difference sets. The parameters can be written as $(4n + \gamma, k, k - n; n)$. The equation $k(k - 1) = \lambda(v - 1)$ gives the quadratic equation

$$k^2 - k = (k - n)(4n + \gamma - 1)$$

which reduces to

$$\left(k - 2n - \frac{\gamma}{2}\right)^2 = n(\gamma + 1) + \frac{\gamma^2}{4}$$

and which gives the desired parameters: In case that $\gamma = 1$, we put $2n + (1/4) =: ((t + 1)/2)^2$ to get the desired parameter representation.   □

This corollary is one of the reasons why people started the investigation of cyclic Hadamard difference sets: Every theorem that rules out the existence of a cyclic Hadamard difference set of order n is a non-existence theorem about a Barker sequence of length $4n$. There are many non-existence results about cyclic Hadamard difference sets. In particular, no single example of a cyclic Hadamard difference set of order $n \neq 1$ is known (the case $n = 1$ and $v = 4$ is the only rather trivial example). This gives strong evidence that no Barker sequence of even length > 4 does exist. Theorem 2.4.11 shows that n has to be odd! We note that there is no theoretical reason why the existence of a cyclic Hadamard difference should imply the existence of a Barker sequence: This is sometimes stated a little bit misleading in the literature. (Presumably there are both no cyclic Hadamard difference sets and no Barker sequences and therefore one exists if and only if the other one exists!)

For a long time, the only non-existence results on Barker sequences have been corollaries from non-existence results on cyclic Hadamard difference sets. As far as I know the first genuine non-existence result on Barker sequences which does not automatically rule out the existence of cyclic Hadamard difference sets is the following:

Result 6.1.6 (Eliahou, Kervaire, Saffari [67]) *Let v be an even integer having a prime factor $p \equiv 3 \bmod 4$. Then there is no Barker sequence of length v.*

Jedwab and Lloyd have used this result together with some non-existence results on cyclic Hadamard difference sets to show the following:

Corollary 6.1.7 (Jedwab, Lloyd [94]) *There is no Barker sequence of length $4u^2$ with $u \leq 5,000$ unless possibly*

$$u \in \{689, 793, 1885, 2525, 3965, 4525\}.$$

We emphasize that this is a result about Barker sequences and <u>not</u> about cyclic Hadamard difference sets. As far as I know, the smallest open case where the existence of a cyclic Hadamard difference set of order u^2 is still undecided is the case $u = 55$.

The challenging open question about Barker sequences and perfect sequences is the following:

Problem 13 Do Barker sequences or perfect sequences of even length > 4 exist?

Although not many periodic Barker sequences of odd length exist, there are some examples of cyclic difference sets where the corresponding ± 1-sequence has constant autocorrelation $\gamma = -1$: Every cyclic Paley-Hadamard difference set of order n has this property. Parametrically, we have the following series of examples:

- Singer difference sets with $n = 2^a$ (see Section 3.1);

- Paley difference sets with $n = (p-1)/4$ (p prime) (the corresponding sequences are called **Legendre sequences**, see Result 2.1.9);

- twin prime power difference sets with $n = (p+1)^2/4$, p and $p+2$ are both primes (Result 2.1.9).

We know that there are not just the Singer difference sets with the classical Singer parameters. Sometimes the Paley difference sets also have the classical parameters. Moreover, the GMW-difference sets (Section 3.2) yield inequivalent examples.

In the examples above, we can of course replace "prime" by "prime power" to get more abelian Paley-Hadamard difference sets. However, the difference sets are not cyclic but elementary abelian.

Only one example of a difference set with parameters $(2t(t+1)+1, t^2, t(t-1)/2; t(t+1)/2)$ is known (the case $t = 2$). For more information about this series, we refer to Eliahou and Kervaire [66] and Broughton [35].

Splitting relative difference sets with $n = 2$ yield examples of perfect sequences with entries $\{0, -1, 1\}$, see Section 3.3. We can also consider complex-valued perfect sequences. We must change the definition of the autocorrelation coefficients a little bit to

$$\delta_s(a) = \sum_{i=0}^{v-1} a_i \overline{a_{i+s}}$$

An example of a complex perfect sequence of length 3 is $(1, \zeta, \zeta)$ where ζ is a complex third root of unity. It can be shown that for any odd length v, there exists a perfect sequence whose entries are v-th roots of unity, see Turyn [168].

6.2 Almost perfect sequences

In the last section, we have seen that there is some evidence that no perfect sequences with entries ± 1 can exist. In particular, only one rather trivial example is known. This is one of the reasons why Wolfmann [176] suggested the investigation of **almost perfect sequences**. These are periodic sequence $a = (a_i)$ of period v with entries ± 1 such that

$$\gamma_s(a) = 0$$

for all $0 < s \leq v-1$ with exactly one exception, say $\gamma_t(a) = x$. Let k denote the number of subscripts i such that $a_i = 1$ $(0 \leq i \leq v-1)$. Similarly to Proposition 6.1.4, we get

$$(A+G)(A^{(-1)}+G) = v + xt + 2(k - (v-k))G + vG = v + xt + (4k-v)G \quad (6.4)$$

in $\mathbf{Z}[G]$ where G is the (multiplicatively written) cyclic group of order v. We note that the product in (6.4) is invariant under the map $g \mapsto g^{-1}$, hence $t = t^{-1}$.

This proves that v is even, and t is the unique involution in G. Let $N = \{1, t\}$ be the unique subgroup of G of order 2. We can rewrite the equation (6.4)

$$\frac{A + G}{2} \frac{A^{(-1)} + G}{2} = \frac{v - x}{4} + \frac{4k - v + x}{4}N + \frac{4k - v}{4}(G - N) \qquad (6.5)$$

which shows that $D := (A + G)/2$ is a relative difference set. We obtain the following theorem:

Theorem 6.2.1 (Bradley, Pott [33]) *An almost perfect sequence of length v exists if and only if $v = 2m$ and there exists a cyclic divisible difference set with parameters*

$$\left(m, 2, k, \frac{4k - 2m + x}{4}, \frac{4k - 2m}{4}\right).$$

Here k is the number of entries $+1$ in a cycle of length v of the almost perfect sequence and x is the exceptional autocorrelation coefficient. □

This theorem shows that precisely the divisible difference sets with $v = 4(k - \lambda_2)$ give rise to almost perfect autocorrelation sequences. We refer the reader to Section 2.3 for examples of these divisible difference sets. If $x = 0$, the almost perfect sequences are perfect. Which sequences give rise to relative difference sets? This is the case if and only if $x = 2m - 4k$. Then the relative difference set D is the extension of an $(m, k, 2k - m)$-difference set of order $m - k$. The complement of this difference set is a difference set of the same order $m - k$ with $m - k$ elements, hence the λ-value of this difference set must be 0. But this is only possible if $m - k = 0$ or $m - k = 1$. In the first case, D is a cyclic $(m, 2, m, m/2)$-difference set. We know that the only example of such a difference set occurs for $m = 2$, see Theorem 4.1.1. The other case is more interesting:

Corollary 6.2.2 *Let (a_i) be an almost perfect sequence of length $2m$. If the number of entries $+1$ in $(a_0, \ldots, a_{2m-1})$ is $m - 1$, then the set of residues*

$$\{i : a_i = 1, 0 \leq i \leq 2m - 1\}$$

modulo $2m$ is a cyclic relative difference set with parameters

$$\left(m, 2, m - 1, \frac{m - 2}{2}\right). \qquad (6.6)$$

□

Examples of these relative difference sets exist whenever $m - 1$ is an odd prime power (Theorem 2.2.12). It is rather interesting and amazing that (parametrically) no other examples are known and that the only known examples arise as projections from the classical affine difference sets. Again, we have quite a few numerical conditions on the existence of relative $(m, 2, m - 1, (m - 2)/2)$-difference set. In the cyclic case, we have the following result:

Result 6.2.3 (Jungnickel, Pott, Reuschling [103]) *A cyclic relative $(m, 2, m - 1, (m - 2)/2)$-difference set with $m \leq 424$ exists if and only if $m - 1$ is a prime power.*

It is apparently rather difficult to decide whether a cyclic relative difference set with $m = 426$ exists or not. In many cases in the proof of Result 6.2.3, there is no "easy" argument to prove non-existence. We had to use multipliers and tried to combine orbits under the multiplier group to construct a difference set. If this is impossible (which we checked by an exhaustive (computer) search) no relative difference set can exist. The case $m = 426$ is difficult since one obtains only 10 multipliers!

Problem 14 Is it true that relative difference sets with parameters (6.6) exist only if $m - 1$ is a prime power? Do inequivalent examples exist? Decide the case $m = 426$.

Relative difference sets with parameters (6.6) have been studied by Delsarte, Goethals and Seidel [58]. They were interested in these objects since the cyclic difference sets give rise to so called *negacirculant conference matrices*. We mention the following result which shows that no splitting relative difference sets with the desired parameters can exist. This can be also interpreted in terms of conference matrices: No group invariant conference matrices exist.

Result 6.2.4 (Jungnickel [98]) *Splitting relative difference sets with parameters (6.6) do not exist in abelian groups.*

The case of almost perfect sequences studied by Wolfman [176] is the case corresponding to our relative difference sets. Wolfmann noticed that all his examples have a nice symmetry property which he called a "miracle configuration". He used these configurations as a basis for a computer search for these almost perfect sequences. He did not notice the connection between the sequences and the (divisible) difference sets.

Using our group ring notation, we define

$$R + Rg^{(-1)} = \sum_{i=0}^{2m-1} b_i i \quad \text{in } \mathbf{Z}[G]$$

where R describes the relative difference set with parameters (6.6). It turns out that the integers b_i have the following properties:

$$b_i \equiv b_{i+m} \bmod 2 \quad (i = 1, \ldots, m - 2),$$
$$b_i + b_{m-i-1} \equiv 1 \bmod 2 \quad (i = 1, \ldots, (m - 2)/2).$$

Moreover, we can achieve

$$b_0 = b_{m-1} = 1, \quad b_m = b_{2m-1} = 0 \quad \text{if } m \equiv 2 \bmod 4$$

and

$$b_0 = b_{2m-1} = 1, \; b_{m-1} = b_m = 0 \quad \text{if } m \equiv 0 \bmod 4.$$

In this case, we say that the sequence contains a **miracle configuration**. We are now going to show that the existence of a miracle configuration follows from the existence of the multiplier $m - 1$. First, we need an easy lemma:

Lemma 6.2.5 *Let R be a cyclic relative $(m, 2, m-1, (m-2)/2)$-difference set in the set of residues modulo $2m$. Then there is a translate R' of R which contains 0 and is fixed by the multiplier $m - 1$, see Theorem 1.3.10. This translate has the property $R' \cap \{m/2, 3m/2\} = \emptyset$.*

Proof. We assume that R is fixed by the multiplier $m - 1$. The projection of R onto the cyclic group $\mathbf{Z}_m$ is $\mathbf{Z}_m \setminus \{s\}$ for some $s \in \{0, \ldots, m - 1\}$. Since R and hence its image in $\mathbf{Z}_m$ is fixed by the multiplier $m - 1$, we have $s = 0$ or $s = m/2$. If $s = m/2$, then 0 or m is an element of R. Then the difference set $R + s$ is a difference set with the desired property.

Now let us assume $0, m \notin R'$. Then we have (w.l.o.g.) $m/2 \in R$. If $m \equiv 0 \bmod 4$, we have $(m - 1)m/2 \neq m/2$ in $\mathbf{Z}_{2m}$ which contradicts the assumption that R is fixed by the multiplier $m - 1$. If $m \equiv 2 \bmod 4$, then $(m-1)m/2 = m/2$ in $\mathbf{Z}_{2m}$ and $R - m/2$ describes a difference set fixed by $m-1$ which contains 0. $\square$

Theorem 6.2.6 (Bradley, Pott [33]) *An almost perfect autocorrelation sequence of period $2m$ which contains exactly $m - 1$ coefficients in a generating cycle is equivalent to a sequence containing a miracle configuration.*

Proof. We only consider the case $m \equiv 2 \bmod 4$, the second case is rather analogous. Let G be the cyclic group of order $2m$ generated by g. We define

$$T := R(1 + g^{-1})$$

and have to prove

$$T + T^{(-1)}g^m = 1 + g^{m-1} + g^m + g^{2m-1} \tag{6.7}$$

and

$$T + T^{(-1)}g^{-1} = G \tag{6.8}$$

in $GF(2)[G]$. We choose the relative difference set R such that R contains no element of the forbidden subgroup N, such that $g \in R$ and R is fixed by the multiplier $m - 1$. This is possible in view of Lemma 6.2.5: It is possible to choose R such that $g^0 \in R$. The elements $g^0, g^{m/2}, g^m$ and $g^{3m/2}$ are fixed by the multiplier $m - 1$ if $m \equiv 2 \bmod 4$. The relative difference sets $Rg^{m/2}$ and $Rg^{3m/2}$ do not meet N, and exactly one of these contains the element g. If we choose this translate R, we get $b_0 = b_{m-1} = 1$ and $b_m = b_{2m-1} = 0$ since $g \in R$ implies $g^{m-1} \in R$. (If $m \equiv 0 \bmod 4$, we have to choose a translate R with the property $R^{(m-1)} = Rg^m$, $R \cap N = \emptyset$, $g \in R$.)

Now we compute

$$R(1 + g^{-1}) + R(1 + g^{-1})g^m = RN + RNg^{-1}.$$

Since $RN = G - N$, we obtain (6.7).

Moreover, we get

$$T + T^{(-1)}g^{-1} = R(1 + g^{-1}) + R^{(-1)}(1 + g^{-1}) = (1 + g^{-1})(R + R^{(-1)}).$$

Observe that the multiplier $m - 1$ has the property $g^{a(m-1)} = g^{-a}$ if a is even and $g^{a(m-1)} = g^{m-a}$ if a is odd. This shows that the coefficient of g^{2i} in $R + R^{(-1)}$ is 0 modulo 2 and the coefficient of g^{2i+1} is 1 (note $g^a \in R \Leftrightarrow g^{m-a} \in R \Leftrightarrow g^{-a} \notin R$ for odd integers a). This proves (6.8). □

Example 6.2.7 We take the relative difference set $R = \{4, 6, 7, 8, 11\}$ in $\mathbf{Z}_{12}$. The translate $R + 9 = \{1, 3, 4, 5, 8\}$ satisfies the properties in Theorem 6.2.1 with $m = 6$. The set T is

$$T = \{0, 1, 2, 5, 7, 8\}$$

and

$$b_0 = b_5 = 1, b_6 = b_{11} = 0,$$
$$b_1 = b_7, b_2 = b_8, b_3 = b_9, b_4 = b_{10},$$
$$b_1 + b_4 = b_1 + b_{10} = b_2 + b_3 = b_2 + b_9 = 1.$$

6.3 Abelian difference set codes

If D is an abelian difference set, then we have called the ideal generated by the group algebra element $\sum_{g \in D} g$ in $K[G]$ the **code** (more precisely, the K-code) of the difference set. As a vector space, this is equivalent to the column space of an incidence matrix of the corresponding incidence structure (over K). Here we mean by "equivalent" that the two vector spaces can be transformed to each other via a permutation of the coordinate positions, i.e., a permutation of the points of the design.

Regarding codes of incidence structures, there are two interesting questions. One question is to determine the minimum weight of the code, i.e., to find the non-zero codeword with the smallest possible weight. (The **weight** of a vector is the number of non-zero entries.) We do not consider this question here. It seems that this is a rather difficult question, see Assmus and Key [24]. In case of projective planes of order n, the answer is completely known: The minimum weight vectors are precisely the multiples of the lines.

Here we will try to find the K-dimensions of abelian difference set codes. The dimension is nothing else than the number of characters $\chi : G \to L$ with $\chi(D) \neq 0$, see Proposition 1.2.16, where L is an extension of K which contains a primitive $|G|$-th root of unity. This is only possible if the characteristic of

K does not divide $|G|$. In many interesting cases, this assumption is true. It turns out that we can restrict ourselves to the case where the characteristic of *K* divides the order of the design. Here we will consider the slightly more general version of divisible designs (the statement that only fields whose characteristic divides the order are "interesting" fields will follow immediately). We need some information about the matrix $\mathbf{AA}^t$ (say $\mathbf{M}$) on the right hand side of (1.2) where *A* is a (points vs. blocks) incidence matrix of a divisible $(m, n, k, \lambda_1, \lambda_2)$-design. For this purpose, it is convenient to think of the matrix $\mathbf{M}$ as a group invariant matrix or a group ring element over your favourite group *G* of order *mn* with a subgroup of order *n*. This is possible, although *A* is not necessarily a group invariant matrix. It is of course best to take an abelian group *G*. The matrix $\mathbf{M}$ corresponds to a group ring element *M* of the form

$$M = a + bN + cG \quad \text{in } \mathbf{Z}[G].$$

We want to study the rank of $\mathbf{M}$ considered as a matrix over a finite field $GF(p)$ (*p* prime), hence $M \in GF(p)[G]$. First of all, note that the eigenvalues of $a + bN + cG$ are

$$a, a + bn, a + bn + cmn \tag{6.9}$$

since

$$(bN + cG)(bN + cG - nb)(bN + cG - nb - cmn) = 0,$$

see also Lemma 1.1.4. Hence, if *p* divides none of the numbers (6.9), then the $GF(p)$-rank of $\mathbf{M}$ is *mn*.

   If $(p, mn) = 1$, we can use character theory to determine the $GF(p)$-rank of $\mathbf{M}$. The character value $a + bn + cmn$ occurs for the principal character, the value $a + bn$ for the $m - 1$ characters which are principal on *N* (and non-principal on *G*), and finally the character value *a* occurs for the remaining $mn - n$ characters. Therefore, we know the multiplicities of all eigenvalues no matter whether some of the numbers in (6.9) are equal or not. The situation becomes slightly more involved if *p* divides *mn* since this "kills" some of the eigenvectors.

Case (a) *p divides n*. The matrix $\mathbf{M}$ has just one eigenvalue *a*. The multiplicity of this eigenvalue *a* is $mn - \text{rank}(bN + cG)$ (the dimension of the eigenspace of *a*) where

$$\text{rank}(bN + cG) = \begin{cases} 0 & \text{if } b = c = 0 \\ 1 & \text{if } b = 0, c \neq 0 \\ m - 1 & \text{if } mc + b = 0, b \neq 0 \\ m & \text{if } mc + b \neq 0, b \neq 0. \end{cases} \tag{6.10}$$

Case (b) *p divides m but not n*. If $b = 0$ there is just one eigenvalue *a* which occurs with multiplicity $mn - 1$ (the eigenspace is the vector space orthogonal to *G* with respect to the standard inner product). If $b \neq 0$, the multiplicity of *a* is $mn - m$ (vectors orthogonal to *N*) and the multiplicity of $a + bn$ is $m - 1$ (the elements of type *NX* orthogonal to *G*) if $c \neq 0$, and it is *m* if $c = 0$ (in this case, we do not need orthogonality to *G*).

   We can use these properties of $a + bN + cG$ to derive some bounds on the rank of the incidence matrix *A* of a divisible design. In this case, we have $a = r - \lambda_1$,

$a + bn = rk - \lambda_2 mn$, and $a + bn + cmn = rk$ (Lemma 1.1.4). We note that

$$\text{rank}(\mathbf{A}) \geq \text{rank}(\mathbf{A}\mathbf{A}^t).$$

Furthermore, if $\mathbf{A}$ is a square matrix, then

$$\text{rank}(\mathbf{A}\mathbf{A}^t) \leq \text{rank}(\mathbf{A}) \leq \frac{mn + \text{rank}(\mathbf{A}\mathbf{A}^t)}{2}. \tag{6.11}$$

To see this, note that the dimension of the kernel of $\mathbf{A}\mathbf{A}^t$ is at most the sum of the dimensions of the kernels of $\mathbf{A}$ and $\mathbf{A}^t$.

We do not want to state all the subcases explicitly which follow from the remarks above. Instead, let us look at some examples. A square divisible $(7, 2, 4, 0, 1)$-design yields $M = 4 - N + G$ with eigenvalues 4, 2 and 16. In characteristic 2, we have only the eigenvalue 0 with multiplicity 8. Then (6.11) gives $6 \leq \text{rank}(\mathbf{A}) \leq 10$. One can check that the divisible design with this parameters constructed in Theorem 3.3.4 has GF(2)-rank 10 as we will see later.

Ko and Ray-Chaudhuri [110] constructed a cyclic divisible $(6, 4, 9, 4, 3)$-design. Here $a = 5$, $b = 1$ and $c = 3$ and the eigenvalues of $\mathbf{M}$ over $\mathbf{Q}$ are 5, 9 and 81. The GF(3)-dimension of the code is between 18 and $(24 + 18)/2$ since the eigenvalue 0 occurs with multiplicity 6. It can be shown that the exact GF(3)-dimension is 19. Similarly, we can bound the GF(5)-dimension by 15 (which is also the exact value).

There exists a divisible design with parameters $(7, 4, 6, 2, 1)$, see Theorem 2.3.10 or Jungnickel [95, Corollary 5.9]. We have $a = 4$, $b = 1$, $c = 1$ and the eigenvalues are 4, 8 and 36. In this case, the GF(2)-rank of an incidence matrix is at most 17 since the eigenvalue 0 occurs with multiplicity 22.

In the special case $n = 1$, we obtain a well-known result on the dimensions of codes of $(m, k, \lambda; n)$-designs for which I do not know whom to attribute it. It is, for instance, contained in Bridges, Hall and Hayden [34]. It shows that we can restrict ourselves to the investigation of GF(p)-codes of $(m, k, \lambda; n)$-designs where p divides the order n (which is the first and easy part of Result 6.3.1, see the remarks above). Bridges, Hall and Hayden have also determined the GF(p)-dimensions <u>exactly</u> if p divides n but p^2 does not.

Result 6.3.1 (Bridges, Hall, Hayden [34]) *Let $\mathcal{D}$ be a symmetric (m, k, λ)-design of order $n := k - \lambda$, and let $\mathcal{C}$ be the associated GF(p)-code. Then*

$$dim(\mathcal{C}) \leq \begin{cases} v/2 & \text{if } p|k \\ (v+1)/2 & \text{if } (p, k) = 1. \end{cases}$$

If n is divisible by p but not by p^2, then

$$dim(\mathcal{C}) = \begin{cases} (v-1)/2 & \text{if } p|k \\ (v+1/2 & \text{if } (p, k) = 1. \end{cases}$$

We are now going to study the codes of some of the abelian difference sets in G described in Section 2.1. In the following, all characters are from the abelian group G into a field of characteristic p that contains "enough" roots of unity. The prime p divides the order n of the difference set under consideration. The standard equation $DD^{(-1)} = n + \lambda G$ in $\mathbf{Z}[G]$ shows that <u>at least</u> one of the character values $\chi(D)$ or $\chi(D^{(-1)})$ is 0 for non-principal characters χ (using $\chi(G) = 0$). We do not change the character values for non-principal characters if we replace D by its complement $G - D$, hence the difference between the ranks of the incidence matrices that belong to D and $G - D$ is at most 1 and the difference is determined by $\chi_0(G)$. This holds even without the assumption that the design has a Singer group!

Let us begin our investigation with the cyclotomic difference sets described in Result 2.1.9 (use the notation introduced there). It is rather easy to determine the GF(p)-dimensions of the codes defined by the Paley difference sets. The Paley difference sets have parameters $(q, (q-1)/2, (q-3)/4); (q+1)/4$. The order of the difference set and the group order are relatively prime, hence we can use character theory to determine the dimensions. We use the notation of Result 2.1.9. We have $C_0 = C_1^{(-1)}$ since -1 is not a square in GF(q), and therefore at least one of $\chi(C_0)$ and $\chi(C_1)$ is 0 if χ is non-principal on GF(q). The equation $C_0 + C_1 = \mathrm{GF}(q) - 0$ shows $\chi(C_0) + \chi(C_1) = -1$ ($\chi \neq \chi_0$), thus <u>at most</u> one of the values $\chi(C_0)$ and $\chi(C_1)$ is 0 provided that $\chi \neq \chi_0$. For the principal character, we get $\chi_0(C_0) = (q-1)/2 \neq 0$ in GF(p) since prime divisors of $(q+1)/4$ do not divide $(q-1)/2$. This shows:

Theorem 6.3.2 (MacWilliams, Mann [130]) *The K-code generated by the Paley difference sets consisting of the quadratic residues in GF(q) with $q \equiv 3 \bmod 4$ has dimension $(q+1)/2$ if the characteristic of K is a prime divisor of $(q+1)/4$.*   □

There is another possibility to see that we have just two character values $\chi(C_0)$ where C_0 is the set of quadratic residues. More generally, let C_0 denote the set of e-th powers in GF(q) ($q = ef + 1$). If

$$\chi : \mathrm{EA}(q) \to K$$

is a non-principal character, then

$$\chi_g : \begin{array}{ccc} \mathrm{EA}(q) & \to & K \\ x & \mapsto & \chi(gx) \end{array}$$

is again a character of EA(q) where the multiplication is carried out in GF(q). Thus we can obtain <u>all</u> characters of EA(q) if we let g run through GF(q)*. If we extend the characters to homomorphisms from the group algebra $K[G]$ into K, we get

$$\chi_g(C_0) = \chi(gC_0) = \chi(C_i)$$

for some $i = 0, \ldots, e - 1$. Therefore, the e-th powers attain just e distinct (non-principal) character values which are $\chi(C_i)$, $i = 0, \ldots, e - 1$. Each of these character values occurs with multiplicity $(q-1)/e = f$.

Now we are going to consider case (b) in Result 2.1.9, i.e., the biquadratic residues. Here $D = C_0$ is a $(4t^2 + 1, t^2, (t^2 - 1)/4; (3t^2 + 1)/4)$-difference set $(q = 4t^2 + 1, e = 4)$. Note that the order of the difference set and the group order are, again, relatively prime. According to the remarks above, we have four character values $\chi(C_0)$, $\chi \neq \chi_0$, namely $\chi(C_0)$, $\chi(C_1)$, $\chi(C_2)$ and $\chi(C_3)$. Because of the bound on the $GF(p)$-rank (see Result 6.3.1), at least two of these values are 0. Assume that three character values are 0: Then $\chi(C_0)\chi(C_1) = 0$ for all non-principal characters χ which implies $C_0 C_1 = c GF(q)$ for a suitable $c \in GF(q)$. We have $c \neq 0$ since the principal character maps C_0 and C_1 to t^2 which is relatively prime to the order of the difference set and hence to the characteristic p of K. But this is impossible since there is no representation of 0 as a sum $c_0 + c_1$ with $c_0 \in C_0$ and $c_1 \in C_1$. To see this, note that the set of squares $C_0 \cup C_2$ is closed under inversion since -1 is, in our case, a square. Thus we obtain the following dimension formula:

Theorem 6.3.3 (Pott [145]) *Let D be the $(4t^2 + 1, t^2, (t^2 - 1)/4; (3t^2 + 1)/4)$-difference set consisting of the biquadratic residues in $GF(4t^2 + 1)$. If the characteristic of K divides the order $n = (3t^2 + 1)/4$, then the dimension of the K-code generated by D is $2t^2 + 1$.* □

In case (c) of Result 2.1.9, the $(4t^2 + 9, t^2 + 3, (t^2 + 3)/4; 3(t^2 + 3)/4)$-difference set is the union of the fourth powers in $GF(q)$ together with 0. Again, there are four character values $\chi(C_0)$ (where χ is a character from $GF(q)$ into a field whose characteristic divides $3(t^2 + 3)/4$). The character values are, as above, $\chi(C_0)$, $\chi(C_1)$, $\chi(C_2)$, and $\chi(C_3)$. Note that the order $n = 3(t^2 + 3)/4$ of the difference set and the group order $4t^2 + 9$ are relatively prime: The only possible common divisor is 3, but then $4t^2 + 9 = 3^{2x}$ would be a square which is true only if $t = 0$, and this is a case not covered by Result 2.1.9 (c). Let us consider the $GF(3)$-rank first. Note that 9 cannot divide n, and that k is not divisible by 3 (otherwise n and $4t^2 + 9$ have the common divisor 3, a contradiction). Thus the $GF(3)$-dimension is $2t^2 + 5$ because of Result 6.3.1. From now on we assume $p \neq 3$. At least two values $\chi(C_i)$ must be -1 since $\chi(C_0 + 0) = \chi(C_0) + 1$ must be 0 for at least $(q - 1)/2$ non-principal characters. We obtain

$$\chi(C_0 + 0)\chi(C_1 + 0) = 0$$

for all non-principal characters χ provided that three of the values $\chi(C_i)$ are -1. We obtain

$$(C_0 + 0)(C_1 + 0) = c GF(q). \tag{6.12}$$

We get $c = 1$ since the zero element has just one representation as a sum of elements in $C_0 \cup \{0\}$ and $C_1 \cup \{0\}$, namely $0 + 0 = 0$. To see this, note that -1 is a square and therefore the set of squares $C_0 \cup C_2$ is closed under inversion (as above). We reach a contradiction by applying the principal character to (6.12): On the lefthand side, we obtain $(t^2 + 3)^2$, on the righthand side we get $4t^2 + 9$. Since p divides $3(t^2 + 3)/4$, this is impossible if $p \neq 3$. We still have to check the value $\chi_0(D)$. Since each prime divisor $p \neq 3$ of n also divides k, we have $\chi_0(D) = 0$ if $p \neq 3$. This proves the following formula:

Theorem 6.3.4 (Pott [145]) *Let D be the $(4t^2 + 9, t^2 + 3, (t^2 + 3)/4; 3(t^2 + 3)/4)$-difference set consisting of the fourth powers in $GF(4t^2 + 9)$ together with 0. If C denotes the K-code of the associated design, then*

$$dim(C) = \begin{cases} 2t^2 + 4 & \text{if } p \neq 3 \\ 2t^2 + 5 & \text{if } p = 3 \end{cases}$$

where the characteristic p of K divides the order $3(t^2 + 3)/4$ of the design. □

The Paley difference sets form an infinite series of cyclotomic difference sets. It is not yet known whether the other series are infinite or not. Hence it might be not too interesting to find dimension formulae for these series. I was not able to find such formulae in the cases (d), (e) and (f).

For case (d), there are only two examples known. There is a $(73, 9, 1; 8)$-difference set which is the Singer difference set of $PG(2, 8)$, and therefore the dimension of the $GF(2)$-code generated by it is 28 by Theorem 3.2.9. The next example is a cyclic $(104411704393, 114243^2, 40391^2; 2^3 \cdot 37 \cdot 499 \cdot 77317)$ difference set. In view of Result 6.3.1, only the $GF(2)$-rank of this difference set is not yet known. (We note that the v-value reported in Storer [165] is not correct.)

We know only one example of type (e), a $(26041, 3256, 407; 7 \cdot 11 \cdot 37)$-difference set where all the dimensions are known. The series (f) includes $(127, 63, 31; 2^5)$-difference sets. The six difference sets with these parameters are listed in Section 3.2. The dimensions of the codes generated by them are (using the ordering in Section 3.2) $8, 64, 22, 36, 36$ and 22.

Another important series of difference sets with $(v, n) = 1$ (where we can apply our character approach) and which presumably yield infinitely many examples are the twin prime power difference sets, see Result 2.1.9, again. We recall that these sets are

$$\{(x, y) : x \in GF(q)^*, y \in GF(q + 2)^*, \ x \text{ and } y \text{ are both squares or non-sqares}\}$$
$$\cup \{(x, 0) : x \in GF(q)\}$$

in $GF(q) \times GF(q+2)$. These difference sets have parameters $(4n - 1, 2n - 1, n - 1)$ of Paley-Hadamard difference sets where $n = (q + 1)^2/4$. This case is interesting since the order of a twin prime power difference set is <u>always</u> a square, hence we can never apply Result 6.3.1.

Theorem 6.3.5 (Pott [145]) *Let D be a twin prime power difference set of order $n = (q+1)^2/4$. If p divides $(q+1)/2$, then the K-rank of the corresponding design is $(q^2 + 1)/4$ where the characteristic of K is p.*

Proof. The character group (characters into a suitable extension field of K) of $GF(q) \times GF(q + 2)$ is just the direct product of the character groups of $GF(q)$ and $GF(q + 2)$. We have to determine the character values $\chi(C_0)$, resp $\chi(C_0')$, and $\chi(C_1)$, resp. $\chi(C_1')$, where C_0, resp. C_0', is the set of squares and C_1, resp. C_1', the set of non-squares in $GF(q)$, resp. $GF(q + 2)$. It is interesting that it is not necessary to determine the character values <u>precisely</u>. We know already that there are just two character values $\chi(C_0)$ if $\chi \neq \chi_0$. These two values are $\chi(C_0) =: a$ and $\chi(C_1) =: b$. Similarly, we define $\chi(C'_0) =: a'$ and $\chi(C'_1) =: b'$. Since $\chi(C_0 + C_1) = \chi(\mathrm{GF}(q) - 0) = -1$, we have $a + b = a' + b' = -1$ in K.

Now we have to consider $(\chi\gamma)(D)$ where χ and γ are characters on $\mathrm{GF}(q)$ and $\mathrm{GF}(q + 2)$. We have nine different types of characters: Those which are principal on $\mathrm{GF}(q)$ and those with $\chi(C_0) = a$ and those with $\chi(C_0) = -1 - a$ must be combined with the respective characters of $\mathrm{GF}(q + 2)$.

If $\chi = \chi_0$ is principal on $\mathrm{GF}(q)$ but γ is non-principal on $\mathrm{GF}(q + 2)$, we obtain

$$
\begin{aligned}
(\chi_0\gamma)(D) &= \chi_0(C_0)\gamma(C'_0) + \chi_0(C_1)\gamma(C'_1) + \chi_0(\mathrm{GF}(q)) \\
&= a'(q - 1)/2 + (-1 - a')(q - 1)/2 + q \\
&= (q + 1)/2 = 0 \quad \text{in } K.
\end{aligned}
$$

Similarly, if γ is principal on $\mathrm{GF}(q + 2)$, we get

$$
(\chi\gamma)(D) = a(q + 1)/2 + (-1 - a)(q + 1)/2 = -(q + 1)/2 = 0.
$$

If both characters are principal, then the character value is simply the cardinality of the difference set which is, in our case, relatively prime to n, hence the character value is different from 0 in K.

If χ and γ are both non-principal, then there are just two character values

$$
(\chi\gamma)(D) = \begin{cases} aa' + (-1 - a)(-1 - a') \\ a(-1 - a') + (-1 - a)a' \end{cases}
$$

(note that $\chi(\mathrm{GF}(q)) = 0$). Both character values occur with multiplicity $(q^2 - 1)/2$. If both values were non-zero, the rank of the incidence matrix would be too large, contradicting Result 6.3.1. If both values were 0, the rank of the design would be 1 which is clearly absurd. Hence there are exactly $(q^2 - 1)/2$ non-principal characters χ of $\mathrm{GF}(q) \times \mathrm{GF}(q + 2)$ with character value $\chi(D) \neq 0$ plus the principal character whose value is non-zero, too. This proves our theorem. □

The twin prime power series was generalized by Whiteman [174] and further by Storer [165]. These constructions yield cyclic difference sets with parameters $(901, 225, 56; 13^2)$ and $(6575588101, 1643897025, 410974256; 13^2 \cdot 37^2 \cdot 73^2)$. As in the other examples of cyclotomic difference sets, it is not known whether these examples belong to an infinite family or not. We note that the $\mathrm{GF}(13)$-dimension of the $(901, 225, 56; 13^2)$-difference set is 417.

For the sake of completeness, we also state the $\mathrm{GF}(p)$-dimensions of the remaining cyclic difference set codes in the range of Baumert's table [26]. We consider only the examples with $(v, p) = 1$:

121A,40,13:	GF(3)-dimension	= 16;
121B,40,13:	GF(3)-dimension	= 31;
121C,40,13:	GF(3)-dimension	= 26;
121D,40,13:	GF(3)-dimension	= 41;
133,33,8	GF(5)-dimension	= 46.

We have used the notation of Baumert[26]. The difference set $(121A, 40, 13)$ is a classical Singer difference set.

It is of some interest to determine the dimension of $(m, k, \lambda; n)$-difference set codes or, in the cyclic case, the linear complexities of difference set sequences. If $(m, n) = 1$ we can use the character theoretic approach. We have done this for many series of difference sets with $(m, n) = 1$. We do not yet have a nice formula for the GMW-difference sets. If the reader checks the known series of difference sets he or she will see that there are no other series with a prime divisor p of n which does not divide m.

We also refer the reader to [36] where the dimensions of certain codes corresponding to strongly regular graphs have been investigated.

A related problem is to <u>characterize</u> difference sets (or, more generally, arbitrary incidence structures) by their dimensions. Let us consider the class of $(4n - 1, 2n - 1, n - 1; n)$-designs. A recent investigation of the codes of these so called **Hadamard designs** is in Assmus and Key [24]. If n is a power of 2, then Hamada and Ohmuri [82] have characterized the Hadamard designs with the smallest possible GF(2)-rank to be the point-hyperplane designs of the projective geometry over GF(2). On the other side of the spectrum, we have the designs coming from the Paley difference sets which are the squares in GF$(4n - 1)$. One cannot expect to characterize $(4n - 1, 2n - 1, n - 1; n)$-designs with the largest possible ranks to be the designs corresponding to the Paley difference sets since the designs exist presumably for all values of n and thus many of them have a large $\overline{\text{GF}(p)}$-rank in view of Result 6.3.1 (whenever the square-free part of n is divisible by some prime). However, it might be possible to characterize the Paley difference sets in the class of all Paley-Hadamard <u>difference sets</u> by their rank. In all known examples, the Paley difference sets are the only Paley-Hadamard difference sets of order n whose code has dimension $2n$ over fields whose characteristic divides n (Theorem 6.3.2). As a first step to prove such a characterization, we can prove the following proposition:

Proposition 6.3.6 (Arasu, Pott [17]) *Let D be an abelian $(v, k, \lambda; n)$-difference set with multiplier 2 in G (in particular, v is odd). If the design associated with D has an incidence matrix of GF(2)-rank $(v + 1)/2$, then $v = 4n - 1$, $k = 2n - 1$ and $\lambda = n - 1$ where n is even and v is a power of some prime p. If G is cyclic, then v is a prime, and D consists of the squares or the non-squares modulo v.*

Proof. The order n of D must be even and k must be odd because of the GF(2)-rank of the incidence matrix (see Result 6.3.1). Since 2 is a multiplier of D, we may assume that D is fixed by 2, thus $D^2 = D^{(2)} = D$ in GF(2)$[G]$ and $\chi(D)^2 = \chi(D)$. This shows that $\chi(D)$ is 0 or 1 (where χ is a character from G into a field of characteristic 2). Since n is even, we have $\chi(D)\chi(D^{(-1)}) = 0$ for all non-principal characters and $\chi_0(D) = 1$ since k is odd. There are exactly $(v + 1)/2$ characters χ such that $\chi(D) \neq 0$ because of the dimension of the difference set code. This implies

$$\chi(D) + \chi(D^{(-1)}) + \chi(1) = \begin{cases} 0 & \text{if } \chi \neq \chi_0 \\ 1 & \text{if } \chi = \chi_0 \end{cases}$$

and thus $D + D^{(-1)} + 1 = G$ in $GF(2)G$ and also in $\mathbf{Z}[G]$. This shows that D is a Paley-Hadamard difference set. The remaining statements of the theorem follow directly from Result 5.2.14. $\qquad \square$

The assumption on the $GF(2)$-rank of the incidence matrix is always satisfied if $n \equiv 2 \bmod 4$. Under this assumption, the theorem above is already in Arasu [2], who did not exploit the connection to antisymmetric difference sets.

Up to now, we have only looked at the $GF(p)$-codes of difference sets if $(v, p) = 1$. Our methods do not work if $p|v$ and $p|n$. We want to discuss now a very special case of relative difference sets in G and their $GF(p)$-dimensions where p divides $|G|$: We want to get some information about the $GF(2)$-ranks of the incidence matrices which correspond to the relative difference sets with $n = 2$ which we have considered in Section 3.3. Using the decomposition lemma (Lemma 3.3.1), we can write these relative difference sets

$$R = A + Bi, \quad A, B \in \mathbf{Z}[H], \quad N = \{1, i\}$$

where N is the forbidden subgroup, and H is the (group theoretic) complement of N in G. The set $D = A \cup B$ is a difference set in H. Let us assume that $|H|$ is odd. We get a group invariant incidence matrix

$$\mathbf{T} = \begin{pmatrix} \mathbf{A} & \mathbf{B} \\ \mathbf{B} & \mathbf{A} \end{pmatrix}$$

where $\mathbf{A}$ and $\mathbf{B}$ are (by abuse of notation) the group invariant matrices corresponding to A and B in $K[H]$. Here K is a field of characteristic 2 such that all characters from H into K exist. The vectors e_χ ($\chi \in G^*$) defined in Section 1.2 are eigenvectors of $\mathbf{A}$ and $\mathbf{B}$ with eigenvalues $\chi(A)$ and $\chi(B)$. The vectors

$$\begin{pmatrix} e_\chi \\ 0 \end{pmatrix}_{\chi \in G^*} \quad \text{and} \quad \begin{pmatrix} e_\chi \\ e_\chi \end{pmatrix}_{\chi \in G^*} \tag{6.13}$$

form a basis of the vector space $K^{2|H|}$. We have

$$\mathbf{T} \cdot \begin{pmatrix} e_\chi \\ 0 \end{pmatrix} = \begin{pmatrix} \chi(A)e_\chi \\ \chi(B)e_\chi \end{pmatrix}_{\chi \in G^*} \quad \text{and} \quad \mathbf{T} \cdot \begin{pmatrix} e_\chi \\ e_\chi \end{pmatrix} = \begin{pmatrix} \chi(A+B)e_\chi \\ \chi(A+B)e_\chi \end{pmatrix}_{\chi \in G^*}.$$

Therefore, the matrix $\mathbf{T}$ relative to the basis (6.13) is

$$\mathbf{T}' = \begin{pmatrix} \mathbf{D}' & 0 \\ \mathbf{B}' & \mathbf{D}' \end{pmatrix}$$

where $\mathbf{D}' = \mathrm{diag}(\ldots, \chi(D), \ldots)$ and $\mathbf{B}' = \mathrm{diag}(\ldots, \chi(B), \ldots)$ are diagonal matrices (note that the characteristic of K is 2). From the matrix $\mathbf{T}'$ we see that the

GF(2)-rank of **T** is twice the number of characters with $\chi(A + B) = \chi(D) \neq 0$ (which is the dimension of the GF(2)-code of the difference set D) plus the number of characters χ such that $\chi(B) \neq 0$ but $\chi(A + B) = 0$. I hope that this observation turns out to be useful in order to decide the lifting problem for the Paley difference sets D.

We can prove that the Paley difference sets consisting of the squares or the non-squares in GF(q) admit no lifting to an abelian relative difference set with $n = 2$, see Arasu, Jungnickel, Ma and Pott [11]. However, the problem is open for the complementary $(q, (q + 1)/2, (q + 1)/4)$-difference sets ($q \equiv 3 \bmod 4$). If the difference set can be extended to a relative $(q, 2, (q + 1)/2, (q + 1)/8)$-difference set R in G, then the λ-value must be even, hence $q \equiv 7 \bmod 8$. We write $R = A + Bi$ with $A, B \subset H$ as above (where $H \cong EA(q)$).

Case (a) $q \equiv 15 \bmod 16$. We have $RR^{(-1)} = 0$ in GF(2)[G]. Therefore, the GF(2)-vector space generated by the rows of an incidence matrix is self-orthogonal with respect to the usual inner product (the matrix corresponding to $R^{(-1)}$ is the transpose of the matrix corresponding to R) and the ideal generated by R in GF(2)[G] has dimension at most q. On the other hand, the discussion above shows that the dimension is at least $q - 1$ since the rank of the incidence matrix corresponding to the complement of the Paley difference set is $(q - 1)/2$. If the rank of the matrix corresponding to R is q, then the ideal is *self-dual*, i.e., the ideal generated by R is the same as the ideal that consists of the vectors orthogonal to R. However, the element H is orthogonal to $R = A + Bi$ since $|A|$ and $|B|$ are both even (see the remarks following Lemma 3.3.1), but the element H is not in the ideal $\langle R \rangle$ since H is not orthogonal to itself ($|H|$ is odd). Thus the GF(2)-rank of R is exactly $q - 1$, and since this is exactly twice the dimension of D, we obtain that $\chi(D) = 0$ implies $\chi(B) = 0$. Therefore, we have a lot of information about the decomposition of D into A and B. I hope that it is possible to exploit this condition, perhaps together with some multiplier arguments. If 2 is a multiplier of $A - B$ in $\mathbf{Z}[H]$ which fixes A and B, then the character values $\chi(A)$, $\chi(B)$ and $\chi(D)$ are just 0 and 1 (note that $\chi(A)^2 = \chi(A^{(2)}) = \chi(A)$ if the calculations are carried out in GF(2)[G]). But even under this assumption, I am not yet able to prove the impossibility of a lifting.

Case (b) $q \equiv 7 \bmod 16$. Now we have $RR^{(-1)} = G - N$ and the dimension of the ideal generated by $G - N$ in GF(2)[G] is $q - 1$, see (6.10). The order of the underlying difference set is congruent 2 modulo 4, and hence we can apply Result 6.3.1.

The elements $E + Fi$ in GF(2)[G] with the properties $E, F \in$ GF(2)[H] (or $E, F \subseteq H$) and $E + F = 0$ or H satisfy

$$(G - N)(E + Fi) = (|E| + |F|)G + (E + F)N = 0 \quad \text{in GF(2)[}G\text{]}$$

since $|E| \equiv |F| + 1 \bmod 2$ if $E + F = H$. These elements form an ideal of dimension $q+1$ which is the annihilator ideal of $\langle G-N \rangle$: Note that the dimension of the **annihilator ideal** ann($\mathcal{I}$) of an ideal $\mathcal{I}$ defined by

$$\text{ann}(\mathcal{I}) := \{X \in K[G] : MX = 0 \text{ for all } X \in \mathcal{I}\}$$

is the same as the dimension of the ideal $\mathcal{I}^{\perp}$ consisting of the elements orthogonal to the subspace $\mathcal{I}$ with respect to the standard inner product. In fact, the two ideals are isomorphic under the isomorphism $X \to X^{(-1)}$ if G is abelian. In order to determine the dimension of $\langle R \rangle$ in GF(2)[G], we can try to determine the dimension of $\langle R \rangle^{\perp}$ or ann($\langle R \rangle$). Since ann($\langle RR^{(-1)} \rangle$) $\supseteq$ ann($\langle R \rangle$), we know where we have to look for elements X with $RX = 0$. We assume that $E + F = H$ and obtain

$$
\begin{aligned}
(A + Bi)(E + Fi) &= (A + Bi)(E + (H + E)i) \\
&= AE + BE + |A|Hi + \\
&\quad + |B|H + AEi + BEi \\
&= DE + H + (DE + H)i
\end{aligned}
$$

in GF(2)[G] (note that $|A|$ and $|B|$ are odd since $q \equiv 7 \bmod 16$). This group algebra element is 0 if and only if $DE = H$, i.e., if $H \in \langle D \rangle$ in GF(2)[H]. But this is not the case since D (and hence every element in $\langle D \rangle$) has even coefficient sum but H has not. Similarly, we obtain in the case that $E + F = 0$

$$(A + Bi)(E + Fi) = (A + Bi)(E + Ei) = DE + DEi.$$

This element is 0 if and only if $DE = 0$, i.e., E is in the annihilator of D in GF(2)[H]. But we know that the dimension of this ideal is

$$q - \dim(\langle D \rangle) = (q + 1)/2$$

since the dimension of the ideal generated by D in GF(2)[H] is $(q-1)/2$ according to Result 6.3.1. These arguments show that

$$\dim(\langle R \rangle) = 2q - \frac{q+1}{2} = \frac{3q-1}{2}$$

in GF(2)[G] where R is a putative lifting of a Paley-Hadamard difference set of order $q \equiv 7 \bmod 16$: We have not used the fact that R is a lifting of a Paley difference set or that q is a prime power. In analogy to the first case, we conclude that $\chi(D) = 0$ implies $\chi(B) \neq 0$. The dimension formula above shows that the $(7, 2, 4, 1)$-difference set generates a GF(2)-code of dimension 10.

Now we want to give a final application of character theory to a problem in finite geometry. Assmus and Key [23] asked the question whether the characteristic vectors of the hyperovals in a desarguesian projective plane of order 2^e generate the orthogonal code $\mathcal{C}^{\perp}$ (with respect to the standar inner product) of the GF(2)-code $\mathcal{C}$ of the plane. For the geometric background, we refer to Hughes and Piper [91], again. A **hyperoval** in a plane of order n is a set of $n + 2$ points no three of which are collinear. It is easy to see that hyperovals can exist only in planes of even order. A line of the plane intersects a hyperoval in 0 or 2 points, thus the characteristic vectors of hyperovals are contained in the orthogonal complement of the GF(2)-code of the plane.

Let us assume that the plane admits an abelian Singer group G. Then the characteristic function of a point set is nothing else than the corresponding group algebra element with 0/1-coefficients. Using the difference set description of the plane, we can give an affirmative answer to the question of Assmus and Key:

Theorem 6.3.7 (Pott [145]) *Let* Π *be a projective plane of even order n admitting an abelian Singer group G. Let C denote the GF(2)-code of* Π *considered as an ideal in GF(2)[G]. Then* $C^\perp$ *contains a hyperoval H such that the ideal generated by H in GF(2)[G] is the orthogonal code* $C^\perp$ *of C (with respect to the standard inner product).*

Proof. Consider a difference set D that describes the plane Π with Singer group G. The equation $DD^{(-1)} = n + G = G$ in GF(2)[G] shows that $\chi(D)\chi(D^{(-1)}) = 0$ for all non-principal characters χ defined in a suitable extension field GF(2^s) of GF(2). It is known and not difficult to see (Jungnickel and Vedder [104]) that $D^{(-1)} \cup \{1\}$ "is" a hyperoval in Π provided that D is fixed by the multiplier 2, i.e., fixed under the group automorphism $x \mapsto x^2$. As usual, we may assume this w.l.o.g. The assumption that D is fixed under the multiplier 2 implies $\chi(D) \in$ GF(2) since $\chi(D) = \chi(D^{(2)}) = \chi(D^2) = (\chi(D))^2$ in GF(2)[G]. Therefore, the number of characters χ with

$$\chi(D^{(-1)} \cup \{1\}) = \chi(D^{(-1)}) + \chi(1) = \chi(D^{(-1)}) + 1 \neq 1$$

is precisely the number of characters with $\chi(D^{(-1)}) = 0$. But this number equals the number of characters with $\chi(D) = 0$ since D and $D^{(-1)}$ describe isomorphic planes. This shows that the dimension of the code generated by the hyperoval $D^{(\neg 1)} \cup \{1\}$ is $n^2 + n + 1$ minus the GF(2)-dimension of the code generated by the difference set D. Therefore the code generated by the hyperoval must be $C^\perp$. □

The only known finite planes with a Singer groups are the (cyclic) desarguesian planes, thus our theorem will be of interest presumably only for desarguesian planes. We have stated the more general version since it might be possible to use it for non-existence proofs (as Jungnickel and Vedder have used the existence of the oval $D^{(-1)}$ to prove the analogue of Result 5.2.12 for planar difference sets of even order).

Corollary 6.3.8 *The orthogonal code of the GF(2)-code of* $PG(2, 2^e)$ *can be generated by hyperovals.* □

This corollary answers the question posed by Assmus and Key. One can show that the oval $D^{(-1)}$ is a conic in $PG(2, 2^e)$, see Jungnickel and Vedder [104]. Since all the conics are projectively equivalent, our proof shows that the orthogonal code of $PG(2, 2^e)$ can even be generated by the characteristic vectors of hyperovals (coming from conics) which are in one orbit under the Singer cycle.

Bibliography

[1] M. Antweiler and A. Bömer, *Complex sequences over GF(p^m) with a two-level autocorrelation function and a large linear span*, IEEE Trans.Inf.Th. **38** (1992), 120–130.

[2] K.T. Arasu *On Wilbrink's theorem*, J.Comb.Th. (A) **44** (1987), 156–158.

[3] K.T. Arasu, *Cyclic affine planes of even order*, Discrete Math. **76** (1989), 177–181.

[4] K.T. Arasu, *Falsity on a conjecture on dicyclic designs*, Utilitas Math. **41** (1992), 253–258.

[5] K.T. Arasu, J.A. Davis and J. Jedwab, *A nonexistence result for abelian Menon difference sets using perfect binary arrays*, Combinatorica (to appear).

[6] K.T. Arasu, J.A. Davis, J. Jedwab and S.K. Sehgal, *New constructions of Menon difference sets*, J.Comb.Th. (A) **64** (1993), 329–336.

[7] K.T. Arasu, J.A. Davis, D. Jungnickel and A. Pott, *Some non-existence results on divisible difference sets*, Combinatorica **11** (1991), 1–8.

[8] K.T. Arasu, J.F. Dillon, D. Jungnickel and A. Pott, *The solution of the Waterloo probelm*, J.Comb.Th. (A) (to appear).

[9] K.T. Arasu, W.H. Haemers, D. Jungnickel and A. Pott, *Matrix constructions of divisible designs*, Linear Algebra and Appl. **153** (1991), 123–133.

[10] K.T. Arasu and D. Jungnickel, *Affine difference sets of even order*, J.Comb.Th. (A) **52** (1989), 188–196.

[11] K.T. Arasu, D. Jungnickel, S.L. Ma and A. Pott, *Relative difference sets with $n = 2$*, Discrete Math. (to appear).

[12] K.T. Arasu, D. Jungnickel and A. Pott, *Divisible difference sets with multiplier -1*, J. of Algebra **133** (1990), 35–62.

[13] K.T. Arasu, D. Jungnickel and A. Pott, *The Mann test for divisible difference sets*, Graphs and Comb. **7** (1991), 209–217.

[14] K.T. Arasu, D. Jungnickel and A. Pott, *Symmetric divisible designs with* $k - \lambda_1 = 1$, Discrete Math. **97** (1991), 25-38.

[15] K.T. Arasu and A. Pott, *Some constructions of group divisible designs with Singer groups*, Discrete Math. **97** (1991), 39-45.

[16] K.T. Arasu and A. Pott, *On quasiregular collineation groups of projective planes*, Designs, Codes and Cryptography **1** (1991), 83-92.

[17] K.T. Arasu and A. Pott, *Cyclic affine planes and Paley difference sets*, Discrete Math. **106/107** (1992), 19-23.

[18] K.T. Arasu and A. Pott, *Variations on the McFarland and Spence constructions of difference sets*, Australasian J.Comb. (in press).

[19] K.T. Arasu and A. Pott, *Impossibility of a certain cyclotomic equation with applications to difference sets* (manuscript).

[20] K.T. Arasu, A. Pott and E. Spence, *A GMW-construction for relative difference sets* (in preparation).

[21] K.T. Arasu and S.K. Sehgal, *Some new difference sets*, J.Comb.Th. (A) (to appear).

[22] K.T. Arasu and Q. Xiang, *Multiplier theorems*, J.Comb.Designs (to appear).

[23] E.F. Assmus and J.D. Key, *Baer subplanes, ovals and unitals*. In: "Proc. IMA Workshop on Coding Theory and Design Theory, Part I" (ed. D.K. Ray-Chaudhuri), pp. 1-8, Springer, Berlin (1990).

[24] E.F. Assmus and J.D. Key: "Designs and their Codes". Cambridge University Press, Cambridge (1992).

[25] E. Bannai and T. Ito: "Algebraic Combinatorics I". Benjamin, New York (1984)

[26] L.D. Baumert: "Cyclic Difference Sets". Springer Lecture Notes **182**, Springer, Berlin (1971).

[27] R. Bacher, "Cyclic difference sets with parameters $(511, 255, 127)$", L'Ens.Math. **40** (1994), 187-192.

[28] G. Berman, *Families of generalized weighing matrices*, Canad.J.Math. **30** (1978), 1016-1028.

[29] T. Beth, D. Jungnickel and H. Lenz: "Design Theory". Cambridge University Press, Cambridge (1986).

[30] A. Beutelspacher and U. Rosenbaum: "Projektive Geometrie". Vieweg, Braunschweig (1992).

[31] R.C. Bose, *An affine analogue of Singer's theorem*, J.Indian Math.Soc. **6** (1942), 1–15.

[32] R.C. Bose and W.S. Connor, *Combinatorial properties of group divisible incomplete block designs*, Ann.Math.Stat. **23** (1952), 367–383.

[33] S.P. Bradley and A. Pott, *Existence and non-existence of almost perfect autocorrelation sequences*, IEEE Trans.Inf.Th. (to appear).

[34] W.G. Bridges, M. Hall and J.L. Hayden, *Codes and Designs*, J. Comb. Th. (A) **31** (1981), 155–174.

[35] W.J. Broughton, *A note on table I of "Barker sequences and difference sets"*, L'Ens.Math. **40** (1994), 105–107.

[36] A.E. Brouwer and C.A. van Eijl, *On the p-rank of the adjacency matrices of strongly regular graphs*, J.Alg.Comb. **1** (1992), 329–346.

[37] R.A. Brualdi: "Introductory Combinatorics". North-Holland, Amsterdam (1977).

[38] R.H. Bruck, *Difference sets in finite groups*, Trans.Amer.Math.Soc. **78** (1955), 464–481.

[39] R.H. Bruck and H.J. Ryser, *The nonexistence of certain finite projective planes*, Canad.J.Math. **1** (1949), 88–93.

[40] P. Camion and H.B. Mann, *Antisymmetric difference sets*, J.Numb.Th. **4** (1972), 266–268.

[41] J.W.S. Cassels: "Rational Quadratic Forms". Academic Press, London (1978).

[42] A.H. Chan and R.A. Games, *On the linear span of binary sequences obtained from q-ary m-sequences*, IEEE Trans.Inf.Th. **36** (1990), 548–552.

[43] W.K. Chan, *Necessary conditions for Menon difference sets*, Designs, Codes and Cryptography **3** (1993), 147–154.

[44] Y.Q. Chen, Q. Xiang and S.K. Sehgal, *An exponent bound on skew Hadamard abelian difference sets*, Designs, Codes and Cryptography (to appear).

[45] U. Cheng, *Exhaustive construction of (255, 127, 63)-cyclic difference sets*, J.Comb.Th. (A) **35** (1983), 115–125.

[46] S. Chowla and H.J. Ryser, *Combinatorial problems*, Canad.J.Math. **2** (1950), 93–99.

[47] W.S. Connor, *Some relations among the blocks of symmetrical group divisible designs*, Ann.Math.Stat. **23** (1952), 602–609.

[48] C.W. Curtis and I. Reiner: "Representations of Finite Groups and Associative Algebras". Interscience, New York (1962).

[49] J.A. Davis, *A note on products of relative difference sets*, Designs, Codes and Cryptography **1** (1991), 117–119.

[50] J.A. Davis, *Construction of relative difference sets in p-groups*, Discrete Math. **103** (1992), 7–15.

[51] J.A. Davis, *Almost difference sets and reversible difference sets*, Arch.Math. (to appear).

[52] J.A. Davis, *Partial difference sets in p-groups*, Arch.Math. (to appear).

[53] J.A. Davis and J. Jedwab, *A survey of Hadamard difference sets*. In: "Groups, Difference sets and the Monster" (eds. K.T. Arasu, J. Dillon, K. Harada, S.K. Sehgal and R. Solomon), deGruyter Verlag, Berlin-New York (in press).

[54] J.A. Davis and J. Jedwab, *A note on new semi-regular divisible difference sets*, Designs, Codes and Cryptography **3** (1993), 379-381.

[55] J.A. Davis and S.K. Sehgal, *Using the Simplex code to construct relative difference sets in 2-groups*, Designs, Codes and Cryptography (to appear).

[56] J.A. Davis and K.W. Smith, *A construction of difference sets in high exponent 2-groups using representation theory*, J.Alg.Comb. (to appear).

[57] W. de Launey, *Square GBRDs over non-abelian groups*, Ars Comb. **27** (1989), 40–49.

[58] P. Delsarte, J.M. Goethals and J.J. Seidel, *Orthogonal matrices with zero diagonal II*, Canad.J.Math. **23** (1971), 816–832.

[59] P. Dembowski: "Finite Geometries". Springer, Berlin (1968).

[60] P. Dembowski and T.G. Ostrom, *Planes of order n with collineation groups of order n^2*, Math.Zeitschrift **103** (1968), 239–258.

[61] P. Dembowski and F. Piper, *Quasiregular collineation groups of finite projective planes*, Math.Zeitschrift **99** (1967), 53–75.

[62] P. Dey and J.L. Hayden, *On symmetric incidence matrices of projective planes*, Designs, Codes and Cryptography (submitted).

[63] J.F. Dillon, *Variations on a scheme of McFarland for noncyclic difference sets*, J.Comb.Th. (A) **40** (1985), 9–21.

[64] J.F. Dillon, *Cyclic difference sets and primitive polynomials*. In: "Finite Fields, Coding Theory and Advances in Communications and Computing" (eds. G.L. Mullen and P.J.S. Shiue), pp. 436–437, Marcel Dekker, New York (1993).

[65] R.B. Dreier and K.W. Smith, *Exhaustive determination of (511, 255, 127)-cyclic difference sets* (manuscript).

[66] S. Eliahou and M. Kervaire, *Barker sequences and difference sets*, L'Ens.Math. **38** (1992), 345–382.

[67] S. Eliahou, M. Kervaire and B. Saffari, *A new restriction on the length of Golay complementary sequences*, J.Comb.Th. (A) **55** (1990), 45–59.

[68] J.E.H. Elliott and A.T. Butson, *Relative difference sets*, Ill.J.Math. **10** (1966), 517–531.

[69] R.A. Games, *The geometry of quadrics and correlations of sequences*, IEEE Trans.Inf.Th. **32** (1986), 423–426.

[70] M.J. Ganley, *On a paper of Dembowski and Ostrom*, Arch.Math. **27** (1976), 93–98.

[71] M.J. Ganley, *Direct product difference sets*, J.Comb.Th. (A) **23** (1977), 321–332.

[72] M.J. Ganley and R.L. McFarland, *On quasiregular collineation groups*, Arch.Math. **26** (1975), 327–331.

[73] M.J. Ganley and E. Spence, *Relative difference sets and quasiregular collineation groups*, J.Comb.Th. (A) **19** (1975), 134–153.

[74] S. Gao and W. Wei, *On non-Abelian group difference sets*, Discrete Math. **112** (1993), 93–102.

[75] D. Gluck, *A note on permutation polynomials and finite geometries*, Discrete Math. **80** (1990), 97–100.

[76] S.W. Golomb: "Shift register sequences (Revised edition)". Aegean Park, Laguna Hills (California) (1982).

[77] B. Gordon, W.H. Mills and L.R. Welch, *Some new difference sets*, Can.J.Math. **14** (1962), 614–625.

[78] D.M. Gordon, *The prime power conjecture is true for* $n \leq 2,000,000$, Electronic J.Comb. **1** (1994).

[79] M. Hall, Jr., *Cyclic projective planes*, Duke Math.J. **14** (1947), 1079–1090.

[80] M. Hall, Jr., *A survey of difference sets*, Proc.Amer.Math.Soc. **7** (1956), 957–986.

[81] M. Hall, Jr. and H.J. Ryser, *Cyclic incidence matrices*, Canad.J.Math. **3** (1951), 495–502.

[82] N. Hamada and H. Ohmori, *On the BIB-designs having the minimum p-rank*. J.Comb.Th.(A) **18** (1975), 131–140.

[83] Y. Hiramine, *A conjecture on affine planes of prime order*, J.Comb.Th. (A) **52** (1989), 44–50.

[84] Y. Hiramine, *Factor sets associated with regular collineation groups*, J. of Algebra **142** (1991), 414–423.

[85] Y. Hiramine, *Planar functions and related group algebras*, J. of Algebra **152** (1992), 135–145.

[86] C.Y. Ho, *On multiplier groups of finite cyclic planes*, J. of Algebra **122** (1989), 250–259.

[87] A.J. Hoffman, *Cyclic affine planes*, Can.J.Math. **4** (1952), 295–301.

[88] D.R. Hughes, *Partial difference sets*, Amer.J.Math. **78** (1956), 650-674.

[89] D.R. Hughes, *Collineation groups and generalized incidence matrices*, Trans.Amer.Math.Soc. **86** (1957), 284–296.

[90] D.R. Hughes *Generalized incidence matrices over group algebras*, Ill.J.Math. **1** (1957), 545–551.

[91] D.R. Hughes and F.C. Piper: "Projective Planes". Springer, New York (1982).

[92] B. Huppert: "Endliche Gruppen I". Springer, New York (1967).

[93] K. Ireland and M. Rosen: "A Classical Introduction to Modern Number Theory". Springer, New York (1982).

[94] J. Jedwab and S. Lloyd, *A note on the nonexistence of Barker sequences*, Designs, Codes and Cryptography **2** (1992), 93–97.

[95] D. Jungnickel, *On automorphism groups of divisible designs*, Can.J.Math **24** (1982), 257–297.

[96] D. Jungnickel, *A note on affine difference sets*, Arch.Math. **47** (1986), 279–280.

[97] D. Jungnickel, *On a theorem of Ganley*, Graphs and Comb. **3** (1987), 141–143.

[98] D. Jungnickel, *On automorphism groups of divisible designs II: group invariant generalised conference matrices*, Arch.Math. **54** (1990), 200–208.

[99] D. Jungnickel, *On affine difference sets*, Sankhya(A) **54** (1992), 219–240.

[100] D. Jungnickel, *Difference sets*. In: "Contemporary Design Theory. A Collection of Surveys" (eds. J.H. Dinitz and D.R. Stinson), pp. 241-324, Wiley, New York (1992).

[101] D. Jungnickel: "Finite Fields: Structure and Arithmetics". BI Wissenschaftsverlag, Mannheim (1993).

[102] D. Jungnickel and A. Pott, *Computational non-existence results for abelian affine difference sets*, Congr.Numer. **68** (1989), 91-98.

[103] D. Jungnickel, A. Pott and D. Reuschling, *On the non-existence of negacirculant conference matrices* (in preparation).

[104] D. Jungnickel and K. Vedder, *On the geometry of planar difference sets*, Europ.J.Comb. **5** (1984), 143-148.

[105] D. Jungnickel and K. Vedder, *Generalized homologies*, Mitt. Math. Sem. Gießen **166** (1984), 103-125.

[106] W.M. Kantor, *Projective planes of type I-4*, Geometriae Dedicata **3** (1974), 335-346.

[107] E.L. Key, *An analysis of the structure and complexity of nonlinear binary sequence generators*, IEEE Trans.Inf.Th. **22** (1976), 732-736.

[108] R.E. Kibler, *A summary of non-cyclic difference sets, $k < 20$*, J.Comb.Th. (A) **25** (1978), 62-67.

[109] A. Klapper, *The vulnerability of geometric sequences based on fields of odd characteristic*, J. of Cryptology **7** (1994), 33-51.

[110] H.P. Ko and D.K. Ray-Chaudhuri, *Multiplier theorems*, J.Comb.Th. (A) **30** (1981), 134-157.

[111] R.G. Kraemer, *Proof of a conjecture on Hadamard 2-groups*, J. Comb. Th. (A) **63** (1993), 1-10.

[112] P.V. Kumar, *On the existence of square dot-matrix patterns having a specified three-valued periodic correlation function*, IEEE Trans.Inf.Th. **34** (1988), 271-277.

[113] P.V. Kumar and W. de Launey, *On circulant generalised Hadamard matrices of prime power order*, Designs, Codes and Cryptography (to appear).

[114] C.W.H. Lam, *On relative difference sets*. In: "Proc. 7th Manitoba Conference on Numerical Mathematics and Computing", pp. 445-474 (1977).

[115] C.W.H. Lam, L.H. Thiel and S. Swiercz, *The non-existence of finite projective planes of order 10*, Canad.J.Math. **41** (1989), 1117-1123.

[116] E.S. Lander: "Symmetric Designs: An Algebraic Approach". London Math.Soc.Lect. Notes **75**, Cambridge University Press, Cambridge (1983).

[117] K.H. Leung and S.L. Ma, *Constructions of partial difference sets and relative difference sets on p-groups*, Bull.Lond.Math.Soc. **22** (1990), 533-539.

[118] K.H. Leung and S.L. Ma, *Partial difference sets with Paley parameters*, Proc.Lond.Math.Soc. (to appear).

[119] K.H. Leung, S.L. Ma and V. Tan, *Abelian divisible difference sets with multiplier -1*, J.Comb.Th. (A) **59** (1992), 51–72.

[120] R. Lidl and H. Niederreiter: "Introduction to Finite Fields and their Applications": Cambridge University Press, Cambridge (1986).

[121] R.A. Liebler and K.W. Smith, *On difference sets in certain 2-groups*. In: "Coding Theory, Design Theory, Group Theory. Proceedings of the Marshall Lall Conference" (eds. D. Jungnickel and S.A. Vanstone), pp. 195–211, Wiley, New York (1993).

[122] H. Lüneburg: "Galoisfelder, Kreisteilungskörper und Schieberegisterfolgen". BI Wissenschaftsverlag, Mannheim (1979).

[123] S.L. Ma. *Partial difference sets*, Discrete Math. **52** (1984), 75–89.

[124] S.L. Ma, *Polynomial addition sets and polynomial digraphs*, Linear Algebra and Appl. **69** (1985), 213–230.

[125] S.L. Ma, *Reversible relative difference sets*, Combinatorica **12** (1992), 425-432.

[126] S.L. Ma, *Regular automorphism groups on partial geometries* (manuscript).

[127] S.L. Ma and A. Pott, *Relative difference sets, planar functions and generalized Hadamard matrices*, J. of Algebra (to appear).

[128] S.L. Ma and B. Schmidt, *The structure of the abelian groups containing McFarland difference sets*, J.Comb.Th. (A) (to appear).

[129] S.L. Ma and B. Schmidt, *On (p^a, p, p^a, p^{a-1})-relative difference sets*, Designs, Codes and Cryptography (to appear).

[130] F.J. MacWilliams and H.B. Mann, *On the p-rank of the design matrix of a difference set*, Inf. and Control **12** (1968), 474–489.

[131] H.B. Mann: "Addition Theorems". Wiley, New York (1965).

[132] H.B. Mann and R.L. McFarland, *On Hadamard difference sets*. In: "A survey of Combinatorial Theory" (ed. J.N. Srivastava), pp. 333-334, American Elsevier, New York (1973).

[133] R.L. McFarland: "On multipliers of abelian difference sets". Ph.D. Dissertation, Ohio State University (1970).

[134] R.L. McFarland, *A family of difference sets in noncyclic groups*, J.Comb.Th. (A) **15** (1973), 1–10.

[135] R.L. McFarland, *Difference sets in abelian groups of order $4p^2$*, Mitt. Math. Sem. Giessen **192** (1989), 1–70.

[136] K. Menon, *Difference sets in abelian groups*, Proc.Amer.Math.Soc. **11** (1960), 368–376.

[137] R.C. Mullin and R.G. Stanton, *Group matrices and balanced weighing designs*, Utilitas Math. **8** (1975), 277–301.

[138] N. Nakagawa, *The non-existence of right cyclic planar functions of degree p^n for $n \geq 2$*, J.Comb.Th. (A) **63** (1993), 55–64.

[139] G. Pickert: "Projektive Ebenen". Springer, New York (1975).

[140] H. Pollatsek (personal communication).

[141] A. Pott, *Applications of the DFT to abelian difference sets*, Arch.Math. **51** (1988), 283–288.

[142] A. Pott, *On abelian difference sets with multiplier -1*, Arch.Math. **53** (1989), 510–512.

[143] A. Pott, *On multiplier theorems.* In: "Proc. IMA Workshop on Coding Theory and Design Theory, Part I" (ed. D.K. Ray-Chaudhuri), pp. 286–289, Springer, New York (1990).

[144] A. Pott, *An affine analogue of Wilbrink's theorem*, J.Comb.Th. (A) **55** (1990), 313–315.

[145] A. Pott, *On abelian difference set codes*, Designs, Codes and Cryptography **2** (1992), 263–271.

[146] A. Pott, *A generalization of a construction of Lenz*, Sankhya **54** (1992), 315–318.

[147] A. Pott, *On the structure of abelian groups admitting divisible difference sets*, J.Comb.Th. (A) **65** (1994), 202–213.

[148] A. Pott, *A survey on relative difference sets.* In: "Groups, Difference sets and the Monster" (eds. K.T. Arasu, J. Dillon, K. Harada, S.K. Sehgal and R. Solomon), deGruyter Verlag, Berlin-New York (in press).

[149] A. Pott, *On projective planes admitting elations and homologies*, Geometriae Dedicata (in press).

[150] A. Pott, *On the linear span of sequences derived from GMW-sequences* (in preparation).

[151] A. Pott, D. Reuschling and B. Schmidt, *On multipliers of affine difference sets*, (submitted).

[152] W. Qiu, *A method for studying the multiplier conjecture and a partial solution to it*, Ars Comb. (to appear).

[153] D. Raghavarao: "Combinatorial Problems in Design of Experiments". Wiley, New York (1971).

[154] L. Rónayi and T. Szönyi, *Planar functions over finite fields*, Combinatorica **9** (1989), 315–320.

[155] P.J. Schellenberg, *A computer construction for balanced orthogonal matrices*. In: "Proc. 6th Southeastern Conference on Combinatorics, Graph Theory and Computing", pp. 513–522 (1975).

[156] B. Schmidt: "Differenzmengen und relative Differenzmengen". Doctoral Thesis, University of Augsburg (1994).

[157] B. Schmidt, *On (p^a, p^b, p^a, p^{a-b})-relative difference sets*, (submitted).

[158] M.P. Schutzenberger, *A nonexistence theorem for an infinite family of symmetrical block designs*, Ann.Eugenics **14** (1949), 286–287.

[159] J. Seberry and M. Yamada, *Hadamard matrices, sequences and block designs*. In: "Contemporary Design Theory. A Collection of Surveys" (eds. J.H. Dinitz and D.R. Stinson), pp. 431-560, Wiley, New York 1992.

[160] M. Simon, J. Omura, R. Scholtz and B. Levitt: "Spread Spectrum Communications, Vol.I", Computer Science Press, Rockville, MD (1985).

[161] J. Singer, *A theorem in finite projective geometry and some applications to number theory*, Trans.Amer.Math.Soc. **43** (1938), 377–385.

[162] K.W. Smith, *A table on non-abelian difference sets*. In: "CRC Handbook of Combinatorial Designs" (eds. C.J. Colbourn and J.H. Dinitz), CRC Press, Boca Raton (to appear).

[163] K.W. Smith, *Nonabelian Hadamard difference sets*, J.Comb.Th. (A) (to appear).

[164] E. Spence, *A family of difference sets*, J.Comb.Th. (A) **22** (1977), 103–106.

[165] T. Storer: "Cyclotomy and Difference Sets". Markham Publishing Comp., Chicago (1967).

[166] K. Takeuchi, *A table of difference sets generating balanced incomplete block designs*, Rev.Inst.Internat.Statist. **30** (1962), 361–366.

[167] R. Turyn, *Character sums and difference sets*, Pac.J.Math. **15** (1965), 319-346.

[168] R. Turyn, *Sequences with small correlations*. In: "Error Correcting Codes" (ed. H.B. Mann), pp. 195–228, Wiley, New York (1968).

[169] R. Turyn and J. Storer, *On binary sequences*, Proc.Amer.Math.Soc. **12** (1961), 394–399.

[170] K. Vedder, *Generalised elations*, Bull.Lond.Math.Soc. **18** (1986), 573–579.

[171] W.D. Wallis, A.P. Street and J.S. Wallis: "Room Squares, Sumfree Sets, Hadamard Matrices". Springer Lecture Notes **292**, Springer, New York (1972).

[172] E. Weiss: "Algebraic Number Theory". McGraw-Hill, New York (1963).

[173] D. Welsh: "Codes and Cryptography". Clarendon Press, Oxford (1988).

[174] A.L. Whiteman, *A family of difference sets*, Ill.J.Math. **6** (1962), 107–121.

[175] H.A. Wilbrink, *A note on planar difference sets*, J.Comb.Th. (A) **38** (1985), 94–95.

[176] J. Wolfman, *Almost perfect autocorrelation sequences*, IEEE Trans. Inf. Th. **38** (1992), 1412–1418.

[177] M.-Y. Xia, *Some infinite classes of special Williamson matrices and difference sets*, J.Comb.Th.(A) **61** (1992), 230–242.

[178] Q. Xiang and Y.Q. Chen, *On Xia's construction of Hadamard difference sets* (submitted).

[179] M. Yamada, *On a relation between a cyclic relative difference set associated with the quadratic extensions of a finite field and the Szekeres difference sets*, Combinatorica **8** (1988), 207–216.

[180] K. Yamamoto, *Decomposition fields of difference sets*, Pac.J.Math. **13**, 337–352.

Index